Thiago Ferreira da Conceição

Corrosion protection of magnesium AZ31 alloy by polymer coatings

Thiago Ferreira da Conceição

Corrosion protection of magnesium AZ31 alloy by polymer coatings

An investigation on the potential of polymers in protecting magnesium sheets from corrosion

Südwestdeutscher Verlag für Hochschulschriften

Impressum/Imprint (nur für Deutschland/only for Germany)
Bibliografische Information der Deutschen Nationalbibliothek: Die Deutsche Nationalbibliothek verzeichnet diese Publikation in der Deutschen Nationalbibliografie; detaillierte bibliografische Daten sind im Internet über http://dnb.d-nb.de abrufbar.
Alle in diesem Buch genannten Marken und Produktnamen unterliegen warenzeichen-, marken- oder patentrechtlichem Schutz bzw. sind Warenzeichen oder eingetragene Warenzeichen der jeweiligen Inhaber. Die Wiedergabe von Marken, Produktnamen, Gebrauchsnamen, Handelsnamen, Warenbezeichnungen u.s.w. in diesem Werk berechtigt auch ohne besondere Kennzeichnung nicht zu der Annahme, dass solche Namen im Sinne der Warenzeichen- und Markenschutzgesetzgebung als frei zu betrachten wären und daher von jedermann benutzt werden dürften.

Coverbild: www.ingimage.com

Verlag: Südwestdeutscher Verlag für Hochschulschriften GmbH & Co. KG
Heinrich-Böcking-Str. 6-8, 66121 Saarbrücken, Deutschland
Telefon +49 681 37 20 271-1, Telefax +49 681 37 20 271-0
Email: info@svh-verlag.de

Herstellung in Deutschland:
Schaltungsdienst Lange o.H.G., Berlin
Books on Demand GmbH, Norderstedt
Reha GmbH, Saarbrücken
Amazon Distribution GmbH, Leipzig
ISBN: 978-3-8381-1865-9

Imprint (only for USA, GB)
Bibliographic information published by the Deutsche Nationalbibliothek: The Deutsche Nationalbibliothek lists this publication in the Deutsche Nationalbibliografie; detailed bibliographic data are available in the Internet at http://dnb.d-nb.de.
Any brand names and product names mentioned in this book are subject to trademark, brand or patent protection and are trademarks or registered trademarks of their respective holders. The use of brand names, product names, common names, trade names, product descriptions etc. even without a particular marking in this works is in no way to be construed to mean that such names may be regarded as unrestricted in respect of trademark and brand protection legislation and could thus be used by anyone.

Cover image: www.ingimage.com

Publisher: Südwestdeutscher Verlag für Hochschulschriften GmbH & Co. KG
Heinrich-Böcking-Str. 6-8, 66121 Saarbrücken, Germany
Phone +49 681 37 20 271-1, Fax +49 681 37 20 271-0
Email: info@svh-verlag.de

Printed in the U.S.A.
Printed in the U.K. by (see last page)
ISBN: 978-3-8381-1865-9

Copyright © 2011 by the author and Südwestdeutscher Verlag für Hochschulschriften GmbH & Co. KG and licensors
All rights reserved. Saarbrücken 2011

Contents

1 – Introduction 5
   1.1 – Corrosion of magnesium alloys 8
   1.2 – Coating for magnesium alloys 13
      1.2.1 – Conversion coatings..13
      1.2.2 – Anodizing or plasma electrolytic oxidation process (PEO 15
      1.2.3 – Polymer coatings..16
         1.2.3.1 – Coating methods 19
         1.2.3.2 – Challenges 21
   1.3 – Measurements and evaluation of corrosion..24

2 – Aim of the work 31

3 – Experimental Part 33
   3.1 – Materials...33
   3.2 – Substrate pre-treatment 33
      3.2.1 - HF treatment...33
      3.2.2 – Acid treatments and mechanical grinding 33
   3.3 – Coating preparation...34
      3.3.1 – Polymer solutions 34
      3.3.2 – Spin coating process 34
      3.3.3 – Dip coating process...34
   3.4 – Coating characterization..35
      3.4.1 – Roughness measurements 35
      3.4.2 - OES analyses..35
      3.4.3- FT-IR investigations 35
      3.4.4 - SEM investigations..36
      3.4.5 - XPS analysis...37
      3.4.6 – Adhesion tests..37
      3.4.7- Thermal analyses 38
   3.5 – Corrosion tests 38
      3.5.1 - Electrochemical analysis..38

3.5.2 – Immersion corrosion test 40

4 – Results 41
4.1 – Pre-treatments 41
4.1.1 - Hydrofluoric acid (HF) treatment 41
4.1.1.1 - Weight change and SEM analyses 41
4.1.1.2 – OES analyses 44
4.1.1.3 – FT-IR and XPS investigations 45
4.1.1.4 – Electrochemical investigations 47
4.1.2 – Grinding and acid cleaning 50
4.2 – Polymer coatings 53
4.2.1 – Spin coated poly (ether imide) [PEI] 53
4.2.1.1 – Coating characterization 53
4.2.1.2 – Electrochemical impedance spectroscopy (EIS) 57
4.2.1.3- Influence of substrate pre-treatment 66
4.2.2 – Dip coated poly(ether imide) 69
4.2.2.1 – Coating characterization 69
4.2.2.2 – Electrochemical impedance spectroscopy 73
4.2.2.3 – Influence of substrate pre-treatment 75
4.2.3 – Spin coated PVDF 85
4.2.4 – Dip coated PVDF 86
4.2.4.1 – Coating characterization 86
4.2.4.2 – Electrochemical impedance spectroscopy 89
4.2.4.3 – Influence of substrate pre-treatment 92
4.2.5 – Spin coating of polyacrylonitrile 97
4.2.5.1 – Coating characterization. 97
4.2.5.2 – Electrochemical impedance spectroscopy 99
4.2.6 – Dip coated polyacrylonitrile 103
4.2.6.1 – Coating characterization. 103
4.2.6.2 – Electrochemical impedance spectroscopy 103
4.2.6.3 – Influence of substrate pre-treatment 107
4.2.6.4 – Tests on simulated body fluid (SBF) solutions 111

5 - Discussion of the results 115

5.1 – Substrate pre-treatments 115
5.1.1 – HF treatment 115
5.1.2 – Acetic and nitric acid cleaning 117
5.2 – Poly(ether imide) coatings 119
5.2.1- The influence of solvent 119
5.2.2 The influence of substrate pre-treatment 123
5.2.3 – Mechanism of coating degradation: Interfacial reactions 127
5.3 – PVDF coatings 133
5.3.1 – Influence of solvent 133
5.3.2 – Effect of substrate pre-treatment 134
5.3.3 – Mechanism of coating degradation 135
5.4 – PAN coatings 143
5.4.1 – Influence of solvent 143
5.4.2 – Influence of substrate pre-treatment 146
5.4.3 – Mechanism of coating degradation 148
5.4.4 – Potential use for biomedical applications 150

6 – Summary and conclusions 153

7 – Acknowledgements 155

1 – Introduction

Magnesium is the eight most abundant element on our planet. It can be found in the Earth's crust (constituting 2% of it) and in seawater (where it is the third most abundant dissolved element) as a component of different minerals [1.1, 1.2]. This alkaline metal was discovered in 1808 by Sir Humphrey Davy by the electrolytic splitting of magnesium oxide but it was first industrially produced only 78 years later [1.2]. From this first industrial production until the second world war the amount of magnesium annual production increased from nearly 10 to 235 000 tons. Its current value is around 500 000 tons and its main application is as an alloying element for aluminium (41%), followed by its use as a structural material (32%), in desulphurization of iron and steels, among others uses (14%)[1.3, 1.4]. Since the beginning of its production, magnesium has drawn the attention of industry to its low density combined with similar mechanical properties to that of metals like aluminium and steel, which enhanced the production of lighter metallic components with similar mechanical strength. On the biomedical field, magnesium appeared as a promising biodegradable implant, due to its interesting corrosion properties.

Table 1.1 shows a comparison between physico-chemical and mechanical properties of these materials and other commonly used metals [1.5]. It can be observed that, while unalloyed magnesium has lower mechanical properties compared to aluminium and iron, the magnesium alloys AZ91D and AZ31 have very competitive yield and ultimate tensile strengths, but with much lower density. They render similar performances with much less weight of material. The notation of magnesium alloys adopted in this study is the most accepted one, created by the American Society for Testing and Materials (ASTM), which is made by taking a letter representing each one of the main alloying elements (in order of concentration) and their respective concentration in wt.%. In this way, the alloy AZ31 has the alloying elements aluminium and zinc in a nominal concentration of 3 and 1 wt. % respectively, while the alloy AZ91 has the same alloying elements but in the respective concentrations of 9 and 1 wt. %. The letter "D" in case of AZ91D represents the stage of development of the alloy, which in the case of AZ91D it corresponds to the following general composition (wt.%): Al 8.3 – 9.7; Zn 0.35 – 1.0; Si (max) 0.10; Mn (max) 0.15; Cu (max) 0.30) Fe (max) 0.005; Ni (max) 0.002; others (max) 0.02. Table 1.2 shows the most common alloying elements for magnesium, their respective notation letter and their influence in general properties.

Table 1.1: Comparison between physico-chemical and mechanical properties of magnesium and its alloys with other usually applied metallic materials [1.2, 1.5].

Material	Density (g cm^{-3})	Melting Point (°C)	Yield tensile strength (YTS)		Ultimate Tensile strength (UTS)	
			Rp (MPa)	YTS/density	Rm (MPa)	UTS/density
Magnesium	1.7	649	21	12	90	53
Aluminium	2.7	660	98	36	118	44
Iron	7.9	1535	130	16	262	33
AZ91D-T6* (die cast)	1.8	Min. 421	160	89	230	128
AZ31	1.8	605- 630	155	86	240	133
Al6082-T6	2.7	555	255	94	300	111

* T6 represents a specific heat treatment of the alloy [1.2].

Due to its low mechanical properties, unalloyed magnesium is rarely applied as a structural material, while the family of AZ alloys represents the majority of the used magnesium products. The AZ magnesium alloys present a good combination of properties, especially when prepared by the high pressure die casting (HPDC) method, as good tensile strength, castability and corrosion resistance. When the aluminium content is higher than 6% (in weight) an intermetallic phase is formed ($Mg_{17}Al_{12}$), which is called of β phase and has better corrosion stability compared to the matrix (α phase). Further, the eutectic composition of the Mg-Al solution has a melting point of 437 °C that considerably improves the alloy castability. The addition of zinc is usually made in a maximal content of 1% to avoid cracking problems during solidification [1.2]. This zinc addition further improves the castability and the corrosion behaviour of the alloys. On the other hand, the AZ alloys show low ductility at room temperature, a common problem in magnesium alloys due to its hexagonal close packed (hcp) structure, which hinders a widespread application of magnesium sheets. Further, this alloy is not suitable for biomedical implants due to evidences of neurological problems related to aluminium [1.6-1.8]. The majority of the magnesium components applied in the automotive industry is prepared by the HPDC method [1.2, 1.9, 1.10]. This method produces components with fine grain structure and excellent surface quality with low impurity levels. The negative aspects of this method are the porosity of the prepared components and the high costs [1.2].

Table 1.2: Most commonly used alloying elements, their notation and description of some positive and negative influences [1.2, 1.9].

Element	Notation	Positive influences	Negative Influence
Aluminium	A	mechanical properties, hardness, corrosion resistance, castability	Porosity, stress corrosion cracking susceptibility
Zinc	Z	Tensile strength, corrosion resistance	-
Copper	C	Ultimate strain	Tensile and compressive strength, corrosion resistance
Yttrium	W	Tensile strength, corrosion resistance, castability.	Liability of cracks
Strontium	J	Mechanical properties, grain refinement	-
Zirconium	K	Tensile strength, ductility, grain refinement	Ultimate strain
Manganese	M	Tensile strength, ductility, corrosion resistance	-
Calcium	X	Creep resistance, grain refinement, castability	Liability of cracks
Rare earths	E	Reduces porosity, high temperature strength and creep resistance.	-
Silicon	S	Compressive strength, hardness	Ultimate strain, castability

The application of magnesium sheets is restricted to few components (inner roof frame, inner door frame) due to its low formability at room temperature and to the low surface quality of the currently produced sheets [1.11, 1.12]. The alloy that is most commonly used for sheet production is AZ31 which shows a good combination of strength and ductility [1.2]. Other wrought components are very seldom applied, as forged road wheels, and requires sophisticated surface treatments and coatings to withstand use conditions [1.13]. The application of these wrought components is limited due to their usual low corrosion resistance. Table 1.3 shows some automobile components currently prepared by magnesium alloys.

Table 1.3: Examples of current application of magnesium alloys in automobiles [1.10-1.15].

Body Structure	Interior	Power train
Wheels	Seat frames	Engine blocks
Engine cradle	Instrument Panel	Gear box housing
Fuel Tank barrier	Steering wheels	Automatic transmission
Inner roof frame	Brackets	Oil Pan
Inner door frames	Air bag housing	Cylinder Head Cover
Mirror housing		
Headlight Retainer		
Radiator Support		

1.1 – Corrosion of magnesium alloys

Magnesium alloys are very promising materials for the transportation sector due to the actual urge in the modern society for new cleaner vehicles which can provide the same comfort and performance of the traditional ones but in a much "greener" and economic manner. The production of lighter vehicles is a very promising way to achieve this goal (a possible decrease in 30% on the CO_2 emission is reported for weight saving [1.9]), and this can be accomplished by the replacement of heavier aluminium and steel components by lighter magnesium ones (this estimation is related to a long-term usage of a vehicle. In a short-term, an increase in CO_2 emission, related to the production of magnesium components, should be considered). Different studies in the literature show that a total weight reduction ranging from 124 to 227 kg can be achieved by the replacement of some aluminium and steel components by their magnesium counterparts, representing an average weight reduction of 10 – 20% [1.10, 1.16]. However, only 5 to 50 kg of magnesium is currently applied in automobiles, and a reduction of 20% in the actual weight would need a magnesium amount of 158 kg [1.10]. One of the main reasons for this low magnesium usage is its low corrosion resistance. Magnesium is the construction material with the highest tendency to oxidize [1.2, 1.17, 1.18]. It has a standard reduction potential, which is measured against a standard hydrogen electrode (SHE), of – 2.37 V_{SHE} whereas aluminium and iron have standard reduction potentials of -1.66 V_{SHE} and -0.44V_{SHE}, respectively. This represents a serious barrier to the widespread application of magnesium as a structural material.

On the other hand, while the corrosion properties of magnesium represent a great problem to the transportation sector, they are very attractive for the preparation of biodegradable medical implants such as bone fixations and stents. The application of these magnesium implants avoids a removal surgery, since that the implant would be gradually degraded and absorbed by the body. The corrosion products of magnesium, shown in equations 1.1 to 1.3, are harmless to the human body, and for that reason, a few years after its commercial production, tests with magnesium made screws, sheets and wires were performed in chirurgical procedures [1.19]. However, a too rapid degradation of some implants was observed, with potential risk of inflammation due to excessive hydrogen production and of loss of mechanical integrity of the implant before healing [1.19-1.21]. The required stability and controlled degradation properties in biological environments for orthopaedic implants are not achieved by any of the currently known magnesium alloys.

While the corrosion of metals like iron and aluminium is mainly influence by oxygen, in case of magnesium and its alloys the critical influence is water and chlorine [1.17]. Very little or no influence of oxygen in the corrosion rate of magnesium is reported. The anodic and cathodic partial reactions of magnesium corrosion are shown in equations 1.1 and 1.2 with the respective potential values (in equation 1.1 the potential is positive since that the oxidation reaction is considered). It can be observed that the net potential of magnesium in water (usually called of corrosion potential (E_{cor}) and/or open circuit potential (OCP)) is -1.54V_{SHE}. In chloride solutions and in the presence of some impurities, the free potential of magnesium AZ alloys is around - 1.67 V_{SHE} while for unalloyed magnesium it is approximately - 1.73 V_{SHE}, the highest value for construction metals in such environments (Figure 1.1).

$$Mg_{(s)} \longrightarrow Mg^{2+}_{(aq)} + 2\bar{e} \qquad \Delta E = +2.37V \qquad \text{equation 1.1}$$

$$2H_2O + 2\bar{e} \longrightarrow H_{2(g)} + 2OH^-_{(aq)} \qquad \Delta E = -0.83V \qquad \text{equation 1.2}$$

$$Mg_{(s)} + 2H_2O \longrightarrow Mg(OH)_{2(s)} + H_{2(g)} \qquad \Delta E = -1.54V \qquad \text{equation 1.3}$$

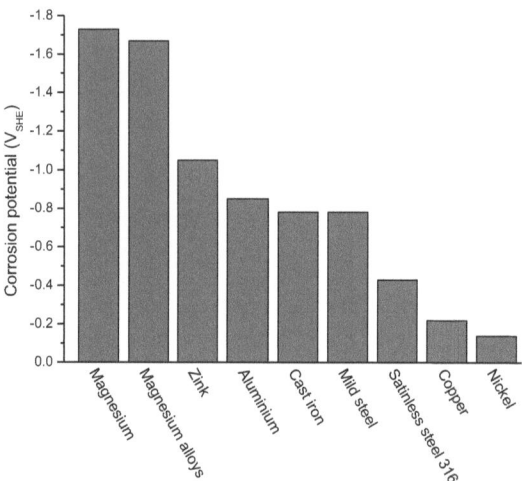

Figure 1.1: Free corrosion potentials of some construction metals in neutral sodium chloride solution [1.2, 1.12, 1.17].

When exposed to atmosphere, clean magnesium samples rapidly become covered by a magnesium oxide/hydroxide film which is generally referred to as "magnesium native film" [1.17]. This native film is partially protective, and for that reason, the atmospheric corrosion of magnesium alloys is good, and can be even better than that of some aluminium alloys [1.2, 1.17]. This native film can be mainly constituted of magnesium oxide or hydroxide depending on the atmospheric humidity. It has very low solubility in water but it is unstable in the presence of anions as Cl^- and SO_2^-, and therefore, cannot provide any protection in such environments. Thus, the corrosion resistance of magnesium in seawater is very low. Only in very basic solutions (pH > 11) magnesium can be stable (in water) since the high concentration of hydroxide renders better stability to the native film [1.2, 1.17].

The corrosion resistance of magnesium alloys strongly depends on the alloying elements, alloy processing and on the impurities level. It was previously commented that the addition of aluminium has beneficial effects in the corrosion behaviour of magnesium due to the formation of a nobler β-phase, which in case it is continuously distributed along the grain boundaries, increases the barrier property of the alloy [1.2]. Nevertheless, some studies in the literature report a negative effect of aluminium addition in the corrosion performance of magnesium [1.22-1.25]. Depending on the volume of the nobler phase, instead of providing a barrier effect it can act as a cathode which accelerates the degradation of the surrounding matrix by a micro-galvanic process [1.26]. This leads to a localized corrosion in chloride environments, which creates pits around the cathodic phase and can lead to the removal of such nobler particles and formation of craters. The scheme shown in figure 1.2 describes this process which is usually known as pitting corrosion. Therefore, the type, size and distribution of secondary phases should be optimized for each alloy system to avoid the formation of micro-galvanic couples. The lower the potential difference between secondary and main phases the lower the micro-galvanic effect [1.17].

Moreover, pitting corrosion can also occur by the influence of impurities [1.2, 1.17]. The cathodic particle shown in figure 1.2 can be either a secondary phase or a metallic particle. Iron, nickel and copper are the most deleterious impurities for magnesium alloys, since that they have low solubility in magnesium and form active cathodic sites [1.17]. As an example, figure 1.3 shows the drastic influence of impurities concentrations in the corrosion rate of magnesium AZ91 alloy. For each alloy there is a tolerance limit content for each impurity, and these values for pure magnesium and some of its alloys are shown in table 1.4 [1.27]. Above this limit, the corrosion rate increases rapidly.

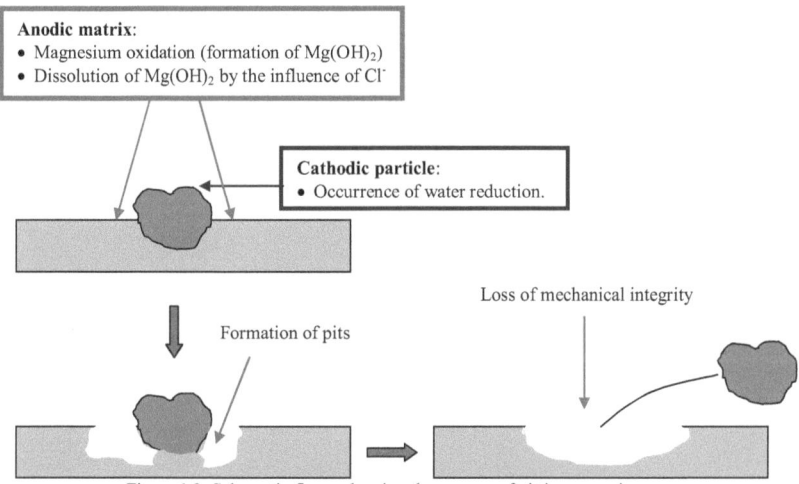

Figure 1.2: Schematic figure showing the process of pitting corrosion.

It can be observed in table 1.4 that the tolerable amount of iron depends on the manganese content for some alloys. It is reported that small amounts of manganese (e.g. 0.2 wt.-%) can considerably improve the corrosion resistance even at iron levels above the tolerance limit [1.2, 1.17]. It is discussed that this positive effect of manganese is related to the formation of intermetallic particles as AlMnFe which has considerably lower cathodic activity than iron. The Fe/Mn ratio is usually referred to as a very important parameter for the corrosion performance of magnesium alloys [1.17]. These examples show that the micro-galvanic effect can be controlled by the proper selection of alloying elements and an optimal distribution of secondary phases. Further information on the tolerance limits for magnesium alloys, its determination and correlation with the microstructure can be found in the studies of Liu et al. and Blawert et al. [1.27, 1.28].

The corrosion rate of magnesium alloys can also be improved by heat and surface treatments [1.26, 1.29]. Heat treatments can change the microstructure of some alloys, especially of those containing aluminium. The proper heat treatment leads to aluminium diffusion from the matrix towards grain boundaries, precipitating as β-phase and improving the corrosion resistance of the alloy [1.26]. Moreover, surface treatments as laser re-melting can considerably improve the corrosion resistance of magnesium alloys by refining the grains and changing secondary phase distribution [1.30-1.32].

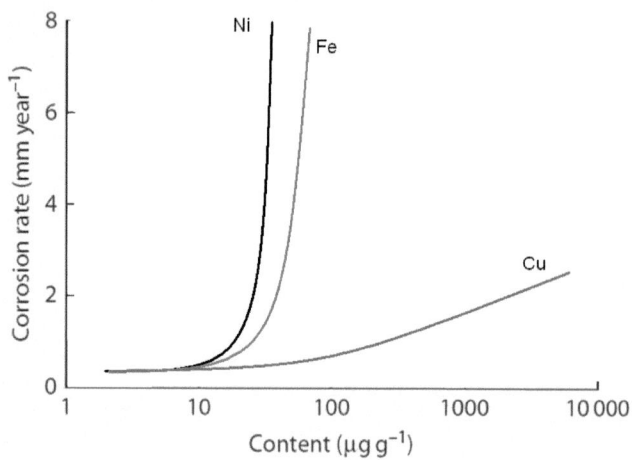

Figure 1.3: Effects of impurities concentrations on the corrosion rate of magnesium alloy AZ91 [1.2].

Table 1.4: Tolerance limits of iron, nickel and copper for magnesium and its alloys [1.27].

Specimen	Fe (ppm)	Ni (ppm)	Cu (ppm)
Pure Mg	170	5	1000
AZ91 (HPDC)	0.032Mn	50	400
AM60	0.021Mn	30	10
AE42	0.020Mn	40	400

In case of magnesium sheets, special attention must be given to the surface quality. The high pressure applied during the rolling process produces deformations at the surface and sub-surface regions of the alloy, which considerably decrease the corrosion resistance [1.17]. The rolling process also deposits metallic impurities on the sheet surface. For that reason the amount of iron on the surface of magnesium sheets is usually much higher than in the bulk. A very cheap and efficient way to reduce this surface impurity level is using acid pickling [1.33, 1.34]. Magnesium dissolves rapidly in all acids (with exception of hydrofluoric and chromic acid, which create protective layers) making acid cleaning a fast and effective way to remove the contaminated layers [1.33, 1.34].

Nevertheless, while acid pickling, heat treatment and microstructure improvement can overcome the micro-galvanic effect, these methods are unable to heal the defects created by the rolling process. Such defects compromise the corrosion resistance of magnesium sheets even at low levels of impurity. Another problem that cannot be solved by these approaches is the macro-galvanic process, which takes place when magnesium gets in contact with steel and aluminium. The macro-galvanic corrosion represents a serious problem for the fastening of magnesium components, as usually used screws are made of iron or aluminium and cause severe corrosion on the magnesium component around the screw [1.13, 1.35]. The only way to enhance the corrosion resistance of magnesium sheets and inhibit the macro-galvanic corrosion is the application of coatings.

1.2 – Coating for magnesium alloys

To enhance the corrosion resistance of magnesium sheets and to avoid galvanic corrosion, magnesium components must be coated in a way that inhibit electric contact between the substrate and the sample surface. This can be performed in many different ways, as described by Gray and Luan in their review on magnesium coatings [1.36]. In this section, the most studied and industrially applied coating methods for magnesium will be discussed, and for a comprehensive review of all possible methods the readers are referred to the publication of Gray and Luan. Before describing the different coating methods it is important to comment that each process must be preceded by a cleaning pre-treatment to remove organic, inorganic and/or metallic impurities that can considerably influence the protectiveness of the coating [1.37]. All coating processes described here can be preceded by cleaning methods as grinding, degreasing and acid pickling.

1.2.1 – Conversion coatings

"Conversion coating" is a term that refers to coating processes where the metal is immersed in a solution which contains certain compounds that react by forming a film. The most common conversion coating processes for magnesium alloys are based on phosphate [1.38-1.40], chromate [1.41], fluorate [1.42-1.48] and stannate [1.49, 1.50] baths, and more recently, on cerium-(IV) baths [1.51, 1.52]. The coatings formed by this process improve the corrosion resistance and can offer wear protection in some degree. Nevertheless, these conversion coatings are more precisely described as pre-treatments since the performance of magnesium alloys coated only by these methods is usually insufficient for a series of applications. This is mostly related to the morphology of the prepared layer, which is usually cracked and porous. Moreover, these coatings provide good adhesion for paints.

Among these conversion coating processes, one that has received considerable attention in the last years is the hydrofluoric acid (HF) treatment. Several researchers reported the use of HF for the treatment of magnesium alloys at different treatment times and acid concentrations [1.42-1.48]. The formed layer is generally described as constituted by MgF_2 and its thickness is usually approx. 2 µm. Considerable attention was given to biomedical orthopaedic implants made of MgF_2 coated magnesium alloy, due to claims that fluoride has a positive influence on bone healing [1.42, 1.43, 1.48]. Even for some industrial applications, where the HF faces serious problems due to its high toxic character, HF has been applied as a pre-treatment for plating and as part of different processes [1.36]. Nevertheless, there is a lack in literature of studies investigating the optimum acid concentration and treatment time, and the reported investigations are usually based on arbitrary choices. Besides that, there is a lack of studies chemically describing the interface of the MgF_2 layer with polymer coatings.

Conversion coatings based on chromate-(VI) are very common for the corrosion protection of magnesium and aluminium alloys. The treatment usually takes only a few minutes and creates a protective layer of approx. 8 µm which provides good protective properties and enhances the adhesion of subsequent coatings [1.41]. Nevertheless, the usage of chromate will be banned soon due to toxicity, and for that reason, different alternatives are been investigated. Table 1.5 shows the main characteristic of the above mentioned conversion coating methods.

Table 1.5: Characteristics of the most used conversion coating processes.

Method	Average thickness (μm)	Morphology and structure	Comments
Phosphate	10	Considerable amount of cracks	Usually provides low corrosion protection but good base for paints
Chromates	10	Porous layer over a non-porous one	Very good corrosion protection at room temperatures.
Fluorates	1.5	Low density of pores	Good corrosion protection. Potential usage in biomedicine
Stannates	3	Globular precipitates	Intermediate corrosion protection
Cerium IV	1.5	Cracks and pores	-

1.2.2 –Plasma electrolytic oxidation process (PEO)

The plasma electrolytic oxidation process (PEO) is the most industrially used method for coating magnesium alloys. Many studies in the literature are dedicated to understand the properties of these coatings on magnesium substrates [1.53-1.58]. This method consists in applying high voltages (usually from 100 to 500V) on a metal piece in an electrolytic bath containing chemicals such as phosphates, silicates, hydroxides, fluorides etc. in variable concentrations. This process forms thick and hard ceramic coatings, which provide good corrosion, abrasion and wear protection. The formed coating is usually described as constituted of different layers with distinct levels of porosity, where the upper layer is the more porous one (figure 1.4). It can have different colours, depending on the constituents, and the thicknesses usually are between 10 and 100 μm.

Three of the most applied and well known PEO processes are patented with the names of KERONITE®, MAGOXID® and TAGNITE®. Different magnesium components currently commercialized are treated by these methods [1.59-1.61]. Nevertheless, due to the upper porous layer, the corrosion protection provided by these processes is not good enough to be used as a single process, and a subsequent sealing procedure is usually required to achieve the necessary performance. Moreover, this method is considerably expensive due to the high required voltages that must be applied during a time of at least about 10 minutes. Besides that, the electrolyte bath should be cooled down by means of a thermostat to avoid excessive temperature increase, what represents additional electrical costs. On the other hand, this method has the advantage of been able to provide protective coatings without any kind of toxic waste (depending on the selected method), is suitable to coat complex shaped substrates and different metals like aluminium and magnesium in only one pass.

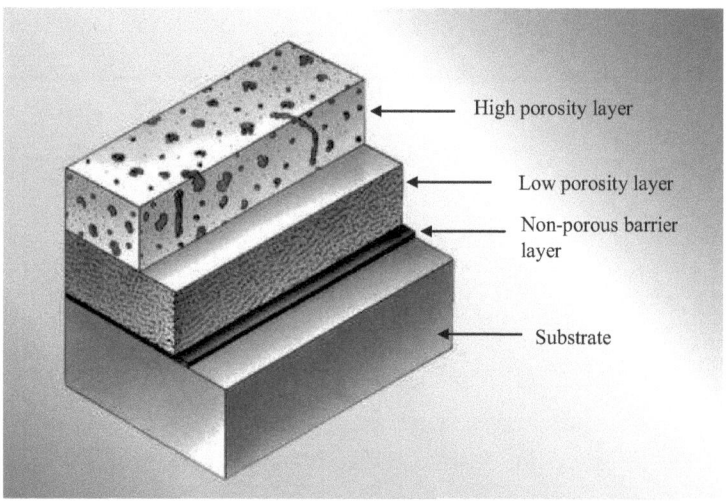

Figure 1.4: Schematic description of a ceramic coating prepared by the PEO process. In this example it is shown the layers of a MAGOXID® coating. (figure from reference 1.59)

1.2.3 – Polymer coatings

Polymers are the matrix component of paints used for all purposes, as in decorative and protective applications [1.62]. The general process of coating a specimen with a polymer is to prepare a solution or an emulsion, apply it to the substrate and let it dry or cure, in case of thermosetting resins. In general the industrial methods of coating metals with polymers are cheaper and easier than those described for conversion and PEO coating (e.g. spraying and dipping) [1.63, 1.64]. In the field of corrosion protection, polymer coatings are usually applied as a sealing process in products previously coated by PEO or conversion coating methods, covering the pores and cracks of these layers. Commercially applied polymer coatings for corrosion protection are usually very thick (from 50 to 100 μm) and constituted of different layers which are classified as: primer, intermediate and top coating. Figure 1.5 shows schematically the general constituents of polymeric coatings as well as their classification according to their nature, corrosion protection mechanism and the description of the function of each one of the different layers [1.65-1.67].

By far and away polymer coatings are the less investigated approach for corrosion protection of magnesium alloys. This can be seen in figure 1.6 which shows the number of

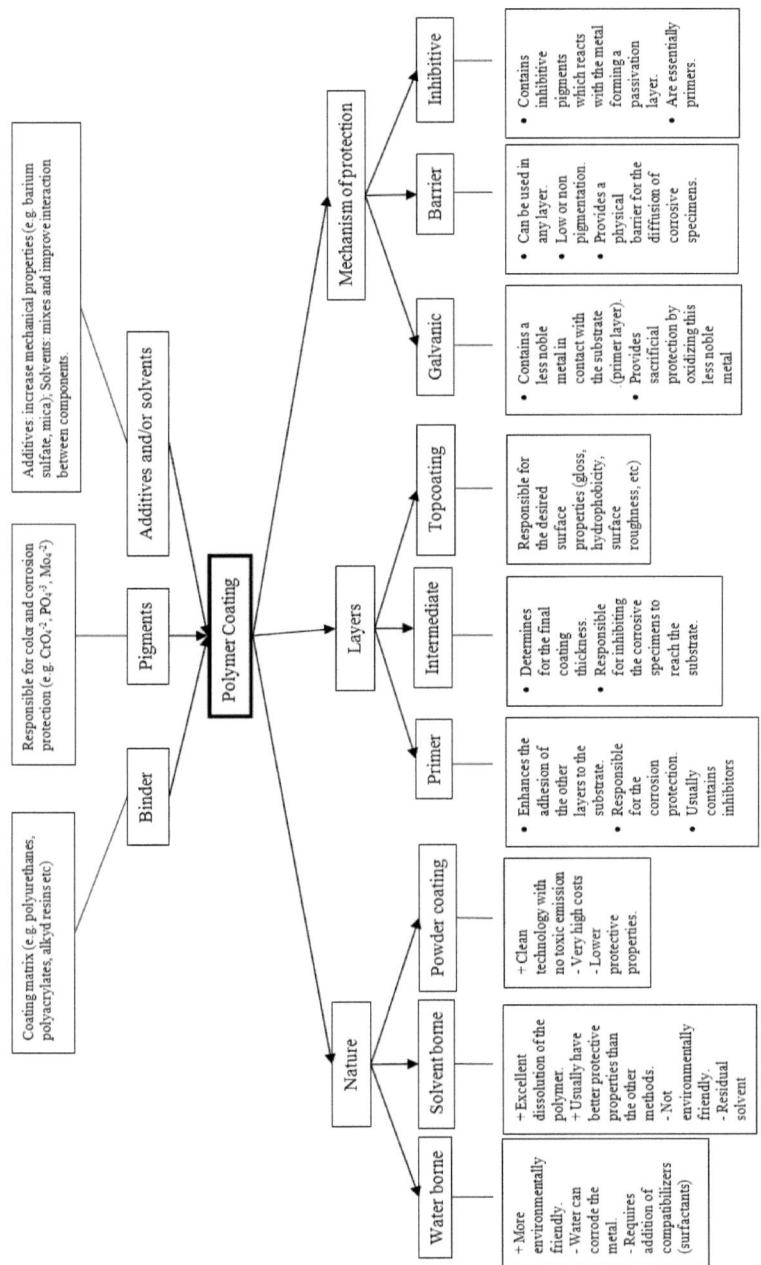

Figure 1.5: Scheme of the constituents and classification of corrosion protection polymer coatings. In the "nature" subsection, the signals "+" and "-"represents advantages and disadvantages respectively [1.65, 1.67].

publications from an inquiry in "ISI-Web of science" using the name of the respective coating process plus "corrosion, magnesium". One possible explanation for this low interest in polymer coatings is that they do not provide good wear and abrasion protection compared to PEO and conversion processes. Besides that, the usual low adhesion of polymers to metals in direct contact may be a factor that drives the attention of researches to other methods.

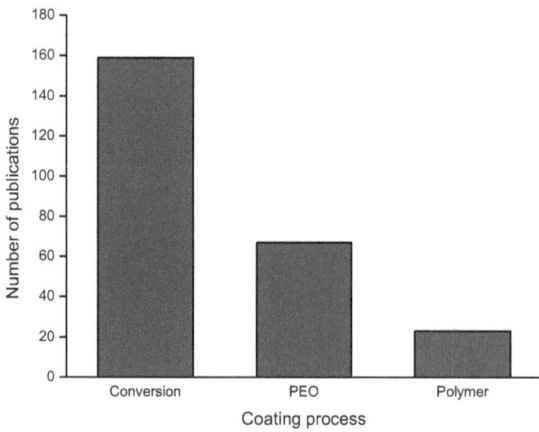

Figure 1.6: Number of publications which resulted from searching in the website "ISIS Web of Science" using the name of the respective coating method plus "corrosion, magnesium".

Nevertheless, polymers has many attractive properties as corrosion protective coatings for magnesium alloys. First of all, polymers can create dense non-porous films with variable thicknesses and high hydrophobicity resulting in highly protective barrier coatings against water and water vapour. Moreover, polymers can be applied on different layers allowing the preparation of multilayered systems. Besides that, with polymer coatings it is possible to control the coating colour by the addition of pigments, a very important aspect for the aesthetical appearance of the coated article, especially for commercial components. Another advantage of polymer coatings is the easiness of the coating methods, since that a simple dipping-drying process can provide thick and protective coatings with minimal consumption of energy.

Another very interesting property of polymer coatings is their high electric resistance. Non-conductive polymers are insulator materials and can have capacitances as low as in the range of 10^{-11} nF cm^{-2} [1.65]. This provides that the substrate will be electrically insulated from the environment and results in very high impedances as reported by Scharnagl et al.[1.68]. Compared to PEO and conversion coating methods, dense polymer coatings provide higher impedances and longer stability in electrochemical tests. However, as previously commented the adhesion of polymer coatings is usually low and leads to coating delamination when water or water vapour reaches the interface. Hence, it is a pre-requisite for a high-performance polymer coating on magnesium alloys that the interface is free from metallic impurities and stable enough to render good adhesion. This can be achieved by the combination of a conversion coating process followed by a polymer coating.

1.2.3.1 – Coating methods

The most simple and often used way to apply polymer coatings to an article is via solution of a specific polymer. This has the negative aspect of generating toxic organic waste but is the method which results in appropriate coating properties. The polymeric solution can be sprayed, brushed, dropped or poured on the substrate and subsequently dried to form the film [1.63, 1.64]. An article can also be dipped into the coating solution for the coating process. Some commonly used methods for polymer coating are shown in table 1.5.

Among these methods, the dip-coating is the most suitable for laboratory studies due to practical reasons. Another method that is adequate for laboratory research is the spin-coating technique, since that spin coaters are available in a variety of sizes. Both methods are suitable for sheet coating and have specific advantages and disadvantages (see table 1.6). The spin-coating method consists in fixating the sample on a chuckle (the sample should be a flat sheet) and spinning it at specific velocities while the polymer solution is dropped on it (figure 1.7). The high spinning speed spreads the solution over the whole sheet surface resulting in thin coatings with good thickness uniformity [1.69-1.75]. The negative characteristic of this process is its limitation to flat substrates and its high sensibility to substrate surface roughness, which induces defects in the coating. Another limitation of the spin-coating method is regarded to the solution viscosity, which should not be too high in order to avoid an uneven spread of coating over the substrate surface. Usually this method results in thickness from below 1 to 5 μm.

Table 1.6: Description of the advantages and disadvantages of some coating methods commonly used in industries [1.63, 1.64].

Coating method	Advantages	Disadvantages
Dip-coating	Simplicity, suitable for substrates with different shapes	Thickness variation during drying
Spray coatings	Simplicity, suitable for any kind of substrate	Poor film control, requires solution with very specific properties
Curtain coating	Speed and control	Break of curtain is possible, too much waste of solution
Spin-coating	Excellent film thickness control, very low waste of solution	Limited to flat substrates, too sensible to substrate surface roughness

The dip-coating method consists in simply dipping an article in the solution, keeping it there for a specific time to allow the wetting of the surface, withdrawing it and letting it dry. The main advantages of the dip-coating method are that it can coat relatively complex shapes, and both sides of sheets simultaneously. Besides that, it can prepare coatings with a variety of thicknesses by varying the solution viscosity (by one single dipping coating process it is possible to prepare coatings with a thickness in the range of 1-100 μm)[1.76-1.78] and is not so sensitive to substrate surface roughness as the spin-coating method. The negative aspect of this method is the non-uniformity of coating thickness along the vertical axis, which takes place during the drying of the sample, as shown in Figure 1.7 [1.64, 1.77].

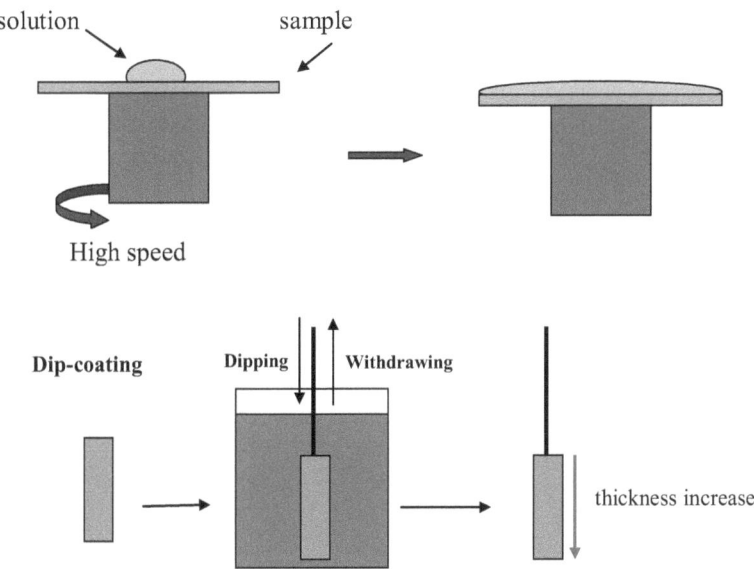

Figure 1.7: Schematic representation of the spin-coating and of the dip-coating methods.

1.2.3.2 – Challenges

Nowadays, there are different challenges that must be overcome for the preparation of highly protective polymer coatings for magnesium alloys. In general, all kinds of polymer coatings (water borne, solvent borne and powder coating) with all kinds of protective mechanism (galvanic, barrier and inhibition) suffers from insufficient adhesion. This requires pre-treatments which produce surfaces capable of synergistically interaction with the polymer, providing higher adhesion and interfacial stability. However, the interface of polymer coatings on magnesium is poorly described in the literature and there is a considerable lack of knowledge about beneficial interfacial interactions. The study of the interface of magnesium alloys with polymers is of great importance in this context.

Moreover, for economic reasons it is important to develop thin and protective coatings. Magnesium components currently applied in industries have very thick coatings, based on many-step processes including conversion, PEO and polymer coatings, as in the method describe by Porsche for corrosion protection of magnesium wheels [1.13]. The thicker the coating the higher the amount of material necessary to protect the metal. This increases not only the price but also the sample weight, reducing the weight saving provided by the light material.

In the field of biomedicine, the challenge is to prepare coatings which provide good corrosion protection in the sense of controlled degradation and biocompatibility which could promote the commercialization of magnesium implants. These coatings should have a high corrosion protection during 2 to 3 months and provide a controlled degradation after that. In this field, polymer coatings are very promising since some polymers can be surface modified for the attachment of bioorganic molecules as proteins and lipids that considerably increases the biocompatibility of the coating.

Many studies in the literature have focused on the preparation of galvanic and inhibitive coatings, while barrier coatings have been less investigated [1.36, 1.79, 1.80]. The development of primers with galvanic protection represents a great challenge, since there are only a few materials with higher tendency to oxidize than magnesium, then being able to provide cathodic protection, as lithium and calcium. Some success was obtained by adding pure magnesium particles in a polymer matrix to act as a primer with galvanic protection for magnesium alloys, as pure magnesium is slightly more anodic than magnesium alloys [1.80]. Nevertheless, the addition of such particles in the matrix considerably decreases its barrier properties. The addition of inhibitive pigments faces similar problems.

There is a considerable need of research on effective barrier coatings for magnesium alloys with beneficial interfacial interaction to the substrate. This field is in focus because the primary protection mechanisms of all coatings are the barrier property and the interface stability. The interface stability is of particular importance because it is impossible to completely avoid the diffusion of water or water vapour through the coating. A stable interface could maintain high corrosion resistance even in the presence of water. It is a much more appropriate approach to previously understand the matrix properties and how it can be optimized and then investigate the influence of additives, rather than preparing galvanic and inhibitive coatings based on arbitrary choices of the matrix. The barrier properties and the interface stability should be the main focus of corrosion protective polymer coatings for magnesium alloys.

To act as a good barrier coating against water, polymers should have a basic property: hydrophobicity. However, polymers that have higher hydrophobicity are usually non-polar, and consequently, have low adhesion to metal substrates. Nevertheless, as previously commented, if an interfacial interaction is present, the adhesion of hydrophobic polymers can be improved. An interesting approach is to coat the metal with polymers that can react with the corrosion product (Mg(OH)$_2$) forming polymer derivatives with higher polarity at the interface. This way, the adhesion would increase inhibiting the corrosion process at the interface, while the top coating would maintain high hydrophobicity. Such interfacial reactions could occur in polymers with functional groups that easily react with bases such as nitrile, ester, imides etc. It is important that this reaction shall not break the polymer chains (decrease in molecular weight), to avoid degradation of the coating, but rather form stable polymer derivatives with higher polarity.

Some polymers that satisfy these criteria are poly(vynilidene fluoride) [PVDF] [1.81-1.83], poly(ether imide) [PEI] [1.84-1.87] and polyacrylonitrile [PAN] [1.88-89], which are commercially available. All these three polymers are hydrophobic and are able to react with bases. The reaction products of these have higher polarity and the reaction does not weaken the polymer chain and stability under environmental conditions. Previous results in the literature show the potential application of PVDF [1.82] and PEI [1.68] as coatings for corrosion protection, while PAN is one of the most interesting polymers for biomedical application, due to its easiness in surface modification [1.89]. The study on the performance of these polymer coatings for magnesium alloys could render significant knowledge about important parameters to achieve good barrier properties and interfacial interaction in polymer coatings. Figure 1.8 shows the chemical structures of these three polymers.

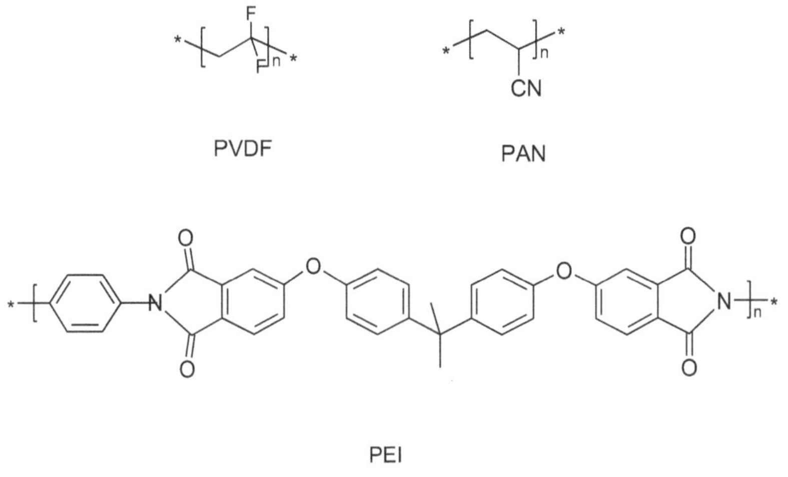

Figure 1.8: Chemical structures of PVDF, PAN and PEI (ULTEM 1000®).

1.3 – Measurements and evaluation of corrosion

The corrosion stability of an uncoated metal is usually described by its corrosion rate, which is a measure of the material weight loss per time and area, and is usually represented as mg/cm^2 day [1.17]. It is also very common to represent the corrosion rate considering the material density at the timescale of one year, resulting the unity mm y^{-1} [1.17]. The traditional non-electrochemical method to evaluate the corrosion rate is by weight loss measurements, where after exposure to corrosion the metal is cleaned with a solution containing chromic acid for removal of corrosion products and weighing for determination of weight loss. This method is applicable to any metallic sample. Other methods like monitoring of gas evolution and determination of ions in solutions are also very common but are only suitable for corrosion processes which produce gases and corrosion products soluble in the corrosive solution, respectively. In case of magnesium alloys, all three methods could be applied when the tests are performed in aqueous chloride solutions.

Electrochemical methods are also very common for the determination of corrosion rates. These are indirect methods where the corrosion current is determined and its correlation to corrosion rate is made considering a previously known corrosion mechanism and the Faraday law [1.90, 1.91]. The corrosion current cannot be directly measured because at OCP all electrons produced in the anodic process are consumed in the cathodic one, and therefore, no net current flows from the system. However, the corrosion current can be determined by polarization methods and the most commonly used one is the direct current (DC) polarization. This method consists in polarizing the natural corrosion potential of a sample by applying a cathodic potential and gradually increasing it towards anodic values [1.90, 1.91]. By extrapolating the tangent (Tafel slopes) of the cathodic and anodic curves to E_{corr} the corrosion current is obtained by the interception of these two curves, as shown in figure 1.9. After the determination of the corrosion current, the corrosion rate can be determined [1.90, 1.91]. As at the corrosion potential the cathodic and anodic currents are the same, the determination of corrosion current can be made using only the cathodic slope. This is of significant importance as the anodic slope is usually non-uniform and difficult to be analysed.

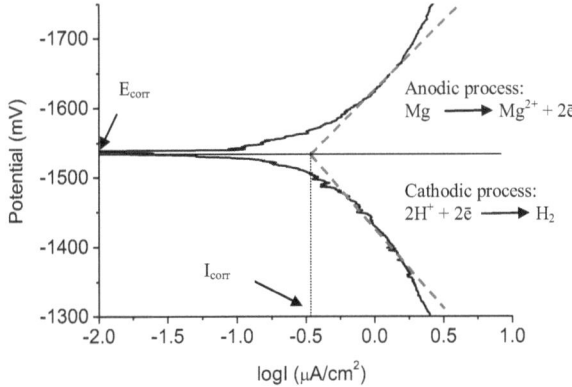

Figure 1.9: Schematic description of the determination of the corrosion rate for Mg from an experimental polarization curve (black solid line) using the Tafel slops (red dashed line).

This polarization technique is widely applied for the determination of corrosion rates of different metals, since it is a simple and fast method. However, the application of this method for corrosion rate determination of magnesium alloys received considerable criticism in the last years due to differences in the results obtained by this and other methods [1.92, 1.93]. Shi et al.[1.92] discuses this subject and shows that Tafel extrapolation does not give reasonable results for magnesium alloys. This is related to the so called "negative difference effect" (NDE) that is regarded as an increase in the cathodic reaction rate even at anodic potentials, which is an unusual and not expected behaviour [1.92, 1.93]. The physico-chemical causes of this phenomenon are still under debate [1.92-1.94]. Nevertheless, the polarization method is still an interesting tool for corrosion analyses of magnesium as it provides correct information on the corrosion potential and corrosion current density which gives insights into the corrosion behaviour of the sample. However, if one wants to discuss corrosion rate of magnesium specimen, methods as weight change and hydrogen evolution measurements should be applied too.

In case of coated magnesium the determination of the corrosion rate becomes difficult because the coating can interfere both in weight loss and in hydrogen evolution measurements. The common approach to study the corrosion performance of a coated magnesium alloy is to investigate the coating stability and the determination when it starts losing its protective properties. After this point, the corrosion rate would be the same as that of an uncoated metal. One of the most used techniques to investigate the stability of coatings in corrosive environments is the electrochemical impedance spectroscopy (EIS) [1.94, 1.67]. The impedance (Z) has the same physical meaning as the resistance (R), with the difference that it varies with the frequency (ω) of the applied potential [1.96]. While in polarization methods a DC potential is applied at a constant rate, in impedance measurements a sinusoidal potential variation is applied at different frequencies, ranging from 10^5 to 10^{-2} Hz. This method allows the determination of the contribution of different elements to the overall sample resistance (impedance), as for example, charge transfer resistance, coating resistance, capacitor resistance, etc. The determination of each one of these electrical elements can be carried out by simulating the impedance spectra using different circuit models [1.67, 1.95, 1.96].

Figure 1.10a and b show two impedance curves for a polymer coating on a magnesium AZ31 alloy with different exposure times to a 3.5 wt.-% NaCl solution. The spectrum that correlates total impedance with the applied frequency is called Bode plot (figure 1.10a), while the one which correlates the real and imaginary parts of Z is called Nyquist plot (figure 1.10b). A plateau in the Bode plot represents a resistance (Z = R when Z does not change with frequency) while the portion of the curve with slope of -1 represents the impedance of a capacitor (the impedance of a capacitor is mathematically defined as: log Z = -log(w) + k, where k is a constant of the material).

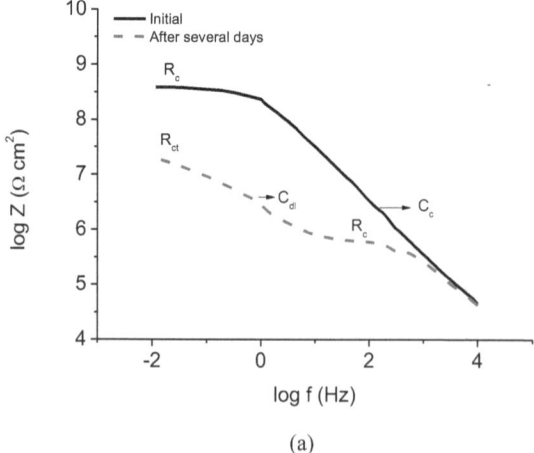

(a)

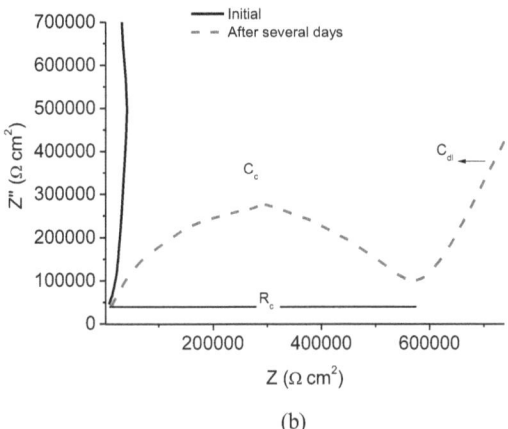

(b)

Figure 1.10: Examples of EIS spectra showing the Bode plot (a) and the Nyquist plot (b) of a sample with different exposure time to the corrosive solution. R_c and R_{ct} represents the coating and charge transfer resistance, respectively while C_c and C_{dl} represents the coating and double layer capacitance, respectively.

After several days of exposure to the corrosive solution, a new plateau (or near plateau) and a new -1 slope appear in the Bode Plot while in the Nyquist plot a semicircle appears (red lines in Figure 1.10). This is related to the concentration of water and ions at the polymer/metal interface which creates an electrochemical double layer. The process of water entering the coating and its concentration increase at the interface, as well as the respective electronic circuits used for the simulation of each condition, is schematically shown in figure 1.11. This new capacitance is usually called "double layer capacitance (C_{dl})".

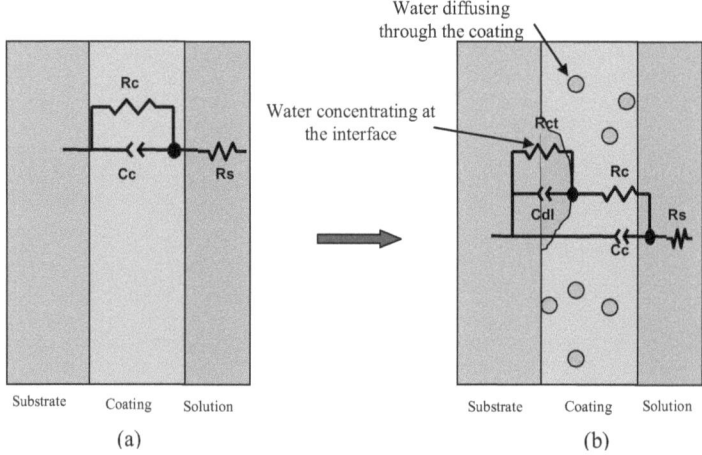

Figure 1.11: Scheme showing the electronic circuits used to simulate the impedance spectra of coated metallic samples: (a) just exposed to the corrosive solution; (b) after several days of exposure to the corrosive solution.

By fitting the impedance spectra using these electronic circuit models it is possible to follow variations in capacitance and resistance with the exposure time and to get information about the stability of the coating. The observation of capacitance variations is particularly important because the capacitance is directly related to the dielectric constant of the coating, as shown in equation 1.4 [1.67, 1.95]. In this equation ε is the dielectric constant of the material, ε_o is the constant of the vacuum, while A and d are the area and thickness of the film, respectively. The dielectric constant of polymers is very sensitive to the presence of water since that water has a much higher dielectric constant (80 while polymers have dielectric constants usually in the range of 2-8) [1.95]. As water diffuses through a coating it produces a capacitance increase, which allows the estimation of water diffusion rates by observation of capacitance variations with time [1.97-1.100]. There are other electronic elements that are also used in the simulation of impedance spectra as inductors and Warburg element [1.96]. Some of these will be briefly described in the results chapter when necessary, as these are not too relevant for the study of polymer coatings.

$$C = \varepsilon \, \varepsilon_o \, A/d \qquad \text{equation 1.4}$$

Another electronic circuit element that frequently appears in simulations of coatings for corrosion protection is called the constant phase element (CPE). This element is capable to describe a resistor, a capacitor, an inductor and elements, which slightly deviate from the pure performance of these. Its impedance is mathematically defined as shown in equation 1.5, where j is the imaginary number, ω is the applied frequency and P and T are the CPE constants. Depending on the values of these constants the CPE can define different elements as follows: when P is equal to 1, the T constant is a pure capacitor; when P is equal to 0, the T constant acts as a resistor; when P is equal to -1, T is an inductor [1.96]. It is possible that the P constant have values different from 1, 0 and -1 and it is in these cases that the substitution of capacitors, resistors and inductors for a CPE becomes important. A much better simulation of real coating systems can be performed using CPEs since deviations in the order of 0.1 in the P constant are normal in corrosion tests.

$$Z = (j\omega)^{-P}/T \qquad \text{equation 1.5}$$

Another very simple method to investigate the corrosion performance of coated samples is the visual observation during exposure to a corrosive environment. For instance, a coated metal sheet could be immersed in a salt solution of specific concentration and composition during a specific time and the formation of corrosion product will be followed. This method does not provide any information about the mechanism of corrosion but it gives insight into the in-service performance of the sample. Besides that, this method allows the determination of edge effects. When the coating is transparent, it is possible to observe when the corrosion products start to form and to correlate this observation with impedance results. Together, impedance and immersion tests are very useful methods to evaluate the protective performance of coated magnesium alloys.

2 – Aim of the work

The aim of the present study is to investigate the potential of polymer coatings for the corrosion protection of magnesium alloys sheets. The influence of parameters such as substrate pre-treatment, solvent type and coating method, on the coating performance will be investigated. The optimal coating conditions for each selected polymer and coating method will be determined. To achieve this aim, the strategy shown in figure 2.1 is adopted. The substrate will be previously cleaned (acid cleaning and grinding), then coated with commercial polymers and finally evaluated in corrosion tests (electrochemical impedance spectroscopy (EIS) and immersion). Additionally, characterization methods like scanning electron microscopy (SEM), Fourier transform infrared spectroscopy (FT-IR), infrared microscopy and x-ray photoelectron spectroscopy (XPS) will support the investigation of the coatings properties. A special attention is given to the determination of the most appropriate conditions for the HF pre-treatment of the alloy. This pre-treatment will be described in details and the performance of the pre-treated samples will be compared to other pre-treatments.

The selected polymers are PEI, PVDF and PAN due to their interesting properties as described in the previous chapter. Especial emphasis will be given to PEI coatings since literature shows a high potential for coatings with this polymer. The conclusions obtained with PEI will be checked for the other polymers aiming to get general and specific conclusions about the coatings performance. At the end of these analyses, the samples with the best and worst performance will be determined as well as the parameters related to this results. The dip-coating and spin-coating methods were selected, both simple and cheap methods suitable for sheet coating.

As substrate, AZ31 Mg alloy is selected which is the most commonly used magnesium alloy for sheet production. The low amount of aluminium renders better ductility for the sheet and increases the biocompatibility of the alloy. Nevertheless it is important to mention that tests made in simulated body fluid (SBF) were performed to give qualitative information about the improvement in corrosion resistance achieved by the used methodology. It is not an aim of the present thesis to use the AZ31 alloy as an implant material due to the mentioned problems associated to aluminium.

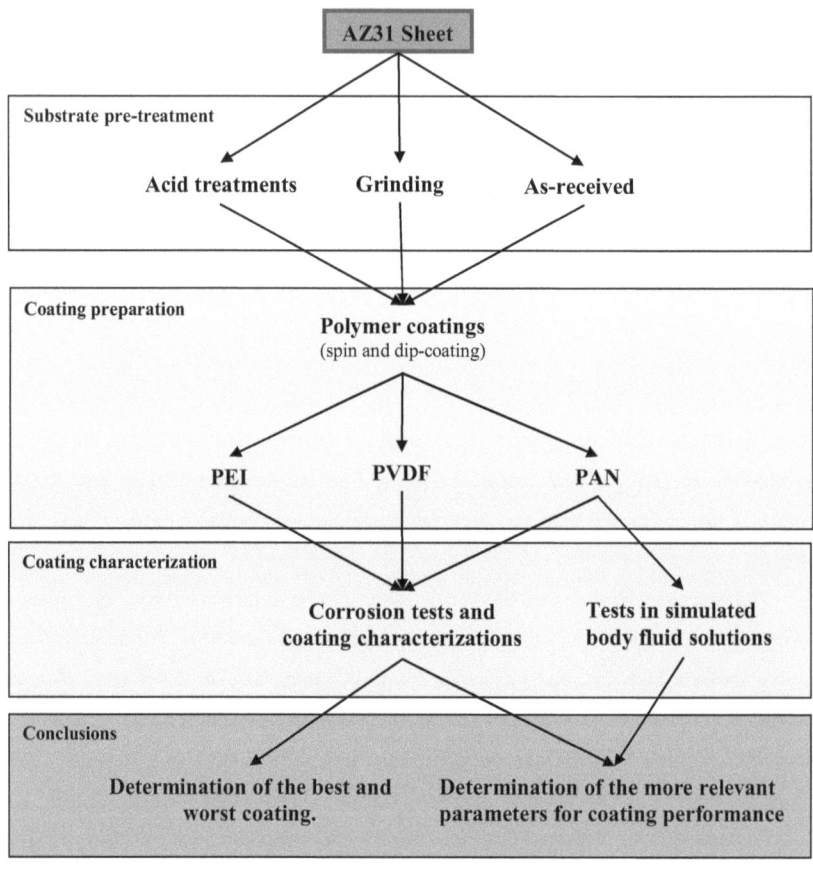

Figure 2.1: General scheme of the experimental strategy adopted in this study.

3 – Experimental Part

3.1 – Materials

Magnesium alloy AZ31 sheets, with chemical composition shown in section 4.1.1 were used as substrate. These sheets were cut in different sizes, ranging from 2 x 2 cm to 5 x 5 cm. The polymers poly (ether imide) Ultem 1000® (Mw: 50.000 g/mol) [PEI] from General Electric, poly (vynilidene fluoride) (Mw: 70.000 g/mol) [PVDF] from Atomchem and polyacrylonitrile (Mw: 130.000 g/mol) [PAN] were used without further purification. The solvents N,N'- dimethylacetamide (DMAc), N-methylpyrrolidone (NMP) and dimethylformamide (DMF), all of synthesis grade were obtained from Merck and used as received. The acids used for the pre-treatment of the substrates (hydrofluoric acid (48% wt), acetic acid (99% wt) and nitric acid (65%) were obtained from Aldrich. A simulated body fluid (SBF) was prepared using the salts NaCl, KCl, $CaCl_2.2H_2O$, $NaHCO_3$ from Merck and $MgSO_4.7H_2O$ and K_2HPO_4 fromChempur, all with a purity level of 99.5 %. The SBF composition will be shown in chapter 4.2.6.

3.2 – Substrate pre-treatment

3.2.1 - HF treatment

The as-received samples were immersed in 80 mL HF in the concentrations of 7, 14, 20 and 28 mol L^{-1} for 1; 5; 15 and 24h, at room temperature. These concentrations and treatment times were selected for practical reasons. The solutions were prepared by dilution of the concentrated 28 mol L^{-1} acid. After the treatment time, the samples were washed with excess of deionised water, dried with non-fuzzing tissue paper to remove water from the surface, and then placed in a vacuum oven (10 mbar) at 40 °C for 1h. The sample weight was measured before and after immersion, using a Mettler Ac 100 analytic balance (± 0.1 mg), to evaluate the weight change. The thickness of the layer formed on the sample's surfaces was measured using a profilometer Hommel Tester T100 performing a scan from a treated to an untreated area of the sample. For this analysis, the samples were not completely immersed in the HF solution and the not immersed part served as reference for the layer thickness determination. Five measurements were performed for each condition. The solution that resulted in the best corrosion protection was also used for samples ground with papers of 800 to 2000 grade to verify the influence of ground surfaces on the corrosion protection.

3.2.2 – Acid treatments and mechanical grinding

Acetic and nitric acid treatments were performed as described in the study of Nwaogu et al.[1.33, 1.34]. The as-received sample was rinsed with ethanol to remove organic impurities at the surface and then dipped in a solution of 5 mol L^{-1} of acetic acid or 1 mol L^{-1} of nitric acid for 2 min. After that the samples were washed with excess of deionised water to remove the acids at the surface and dried in a vacuum oven. The mechanical grinding process consisted in grind the samples using papers from 500 to 2500 grit. The ground surface was finally rinsed using deionised water and the samples stored in clean conditions until required.

3.3 – Coating preparation

3.3.1 – Polymer solutions

Solutions of the polymers were prepared by dissolving the polymer in the appropriate solvent at 80 °C and stirring over night. The concentrations for PEI and PVDF solutions were 10, 15 and 20 wt.-% while for PAN it was 6 and 8 wt.-%. The viscosity of the solutions was determined using a Brookfield R/S-CPS Rheometer. Ten measurements were performed for each solution in the shear rate range of 50-500 s^{-1}. All solutions showed Newtonian behaviour at the applied shear rate.

3.3.2 – Spin-coating process

The spin-coating process was performed in a spin coater Cee$_{TM}$ 200 operated under room or N$_2$ atmosphere. Samples of dimensions 2 x 2 cm or 5 x 5 cm (the last one used specifically for the adhesion tests) were spun at a specific velocity (1000 – 1600 rpm) during 100 s when 3 mL of the polymer solution was applied to the substrate. Prior to the coating process all substrates were rinsed using ethanol. After the coating step, the spin velocity was set to 3000 rpm during 150 s for the drying process. In some specific case (as in the case of PEI coatings prepared using NMP solutions) a second drying process was performed at 3500 rpm during 150 s. This second drying process was necessary to ensure the dryness of the coating, which was not complete after the first one due to the low vapour pressure of NMP at room temperature. The drying of all samples was finalized by storing these under clean conditions for another 20 h at room temperature. Besides that, some samples were also dried in a vacuum oven at 135 °C for 12 h to investigate the influence of residual solvent in the coating performance.

3.3.3 – Dip-coating process

For the dip-coating process, all substrates were pre-heated over a heating plate at 180 °C during 10 minutes to eliminate entrapped air and moisture from the surface. After this pre-heating process, the samples were immersed into the polymer solution during 20 seconds to allow wetting of the surface, and then withdrawn from it to dry. The drying process consisted in hanging the coated sheet in a vacuum oven (10 mbar) at 115 °C (for PEI and PAN) or 150 °C (PVDF) during 12h. These different drying temperatures were selected based on their effects on the coating morphology as will be explained later on the chapter about PVDF coatings. As the samples hung in the vertical position, there was an outflow of solution from the substrate that could affect the thickness uniformity. However, previous tests showed that the thickness uniformity is better when the sample is dried in the vertical than in the horizontal position. The coating thickness uniformity was evaluated using the thickness measurement gauge from Minitest and profilometer measurements.

3.4 – Coating characterization

3.4.1 – Roughness measurements

The surface roughness (R_a) of all substrates (pre-treated and as-received samples) was measured using the profilometer Hommel Tester T100. For each sample, three to five measurements were performed in a scanning range of 4.8 mm. The results presented are an average of these.

3.4.2 - OES analyses

The concentration of impurities and alloying elements on the substrate surfaces was evaluated using optical emission spectroscopy (OES). The analyses were performed in a spectrometer Spectrolab M9, model 2003. The results shown in section 3.1.1 represent an average of three measurements each, performed at different points of the sample surfaces.

3.4.3- FT-IR investigations

To investigate the compounds formed on the substrate surface by the HF treatment, as well as by the corrosion process, and to characterize the conformation of the polymers and crystalline phases present in the coatings, Fourier transform infrared spectroscopy was used. The analyses were performed on a Bruker Tensor 27 IR-spectrometer. The surface of the samples was analyzed using a reflectance unit at an angle of 80 degrees with 2048 scans at a resolution of 4 cm^{-1} in the frequency range of 300 cm^{-1} and 5000 cm^{-1}.

Connected to the infrared spectrometer was an infrared microscope HYPERION 2000. This microscope was used to investigate specific points at the samples surfaces after the corrosion process. The used objective had a magnification power of 15x. These analyses were performed on the visual-reflectance mode using 120 scans at a resolution of 4 cm^{-1}. An image of the used facility is shown in Figure 3.1.

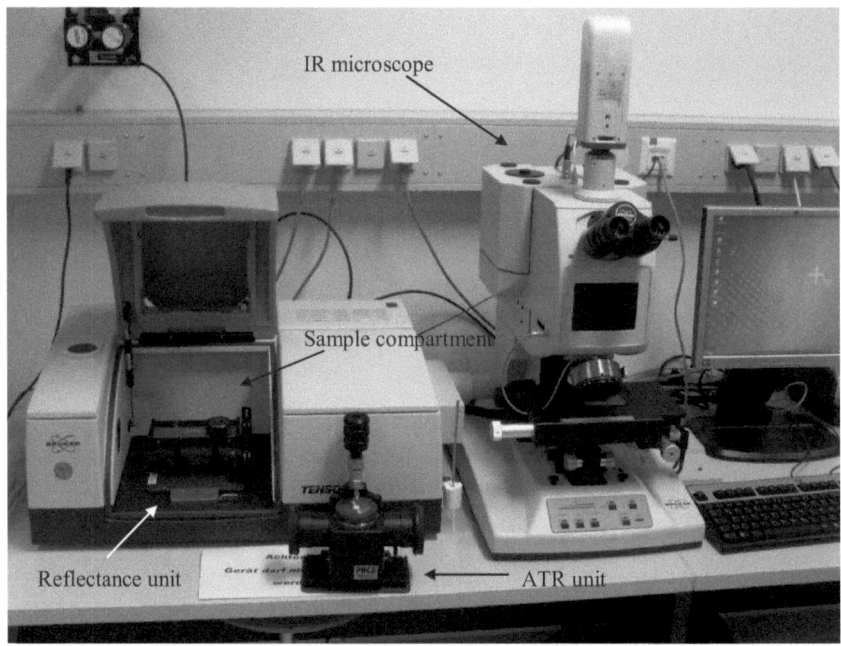

Figure 3.1 Image of the used infrared facility.

3.4.4 - SEM investigations

The morphology of the surfaces was studied using a scanning electron microscope (SEM) Cambridge Stereoscan 200 with an acceleration voltage ranging from 5 to 10 kV. All substrates, including the HF treated one, could be analyzed without gold sputtering due to sufficient surface conductivity. However, all the polymer coatings required prior sputter. The cross section of the prepared coatings was investigated by removing the coating from the substrate, breaking it in liquid nitrogen and fixating the coating on an appropriated support. These procedure was selected instead the grinding of the coated substrate edge to allow the visualization of channels in the coating that could be covered during the grinding process.

3.4.5 - XPS analysis.

X-ray photoelectron spectroscopy (XPS) analyzes were performed in a Kratos DLD Ultra Spectrometer using an Al-K_α X-ray source (monochromator) as anode. For the survey spectra as well as for the region scans a pass-energy of 160 eV was used. The area of interest was limited to 55 μm by an aperture in all cases. Charge neutralization was used for the analyses of all polymer coatings. The concentration and the chemical state of the elements were investigated. The total integral of the XPS intensities (peak area) was used for determining the chemical composition while a linear background subtraction was performed. Depth profiling was carried out by using argon sputtering with energy of 3.8 keV and a current density of 195 μA/cm². The etching rate was calibrated to 36 nm/min using Ta_2O_5.

3.4.6 – Adhesion tests

The adhesion of the coatings to the substrates was evaluated by pull-off test performed on a PosiTest Pull-OFF Adhesion Tester from DeFelsko, in accordance with ASTM D 4541 and ISO 4624. A dolly of 20 mm size was adhered to the coating surface using Alderite adhesive. The analyzed area was isolated by cutting the coating around the dolly using a special tool. The dolly was then connected to the actuator of a hydraulic pump and the strength necessary to pull off the coatings was measured within a resolution of 0.01 MPa. The measurements were performed in dry and wet coatings, where the wet condition was after 12 h of immersion in distilled water. Three to five measurements were performed for each sample. In Figure 3.2 it can be seen the whole equipment with the appropriated tools used for this characterization.

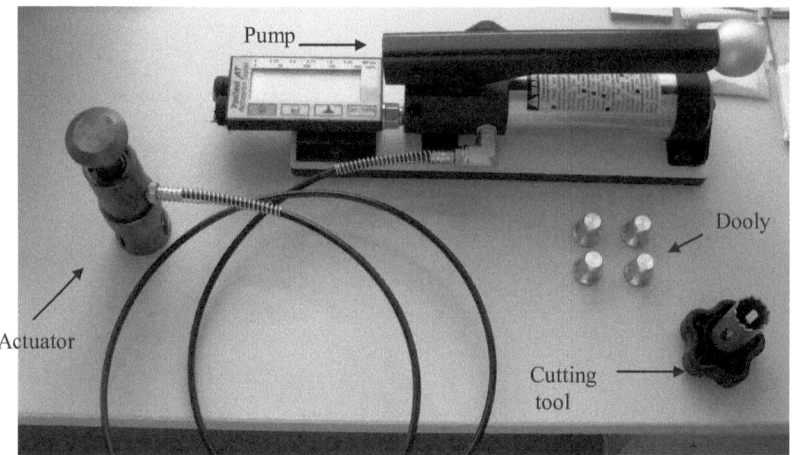

Figure 3.2 Equipment and tools used for the adhesion tests.

3.4.7- Thermal analyses

The determination of the T_g and melt temperature of the polymers was performed using differential scanning calorimetry (DSC) analyses in the NETZSCH DSC 204 equipment. Different procedures were performed depending on the coating and on the aim of the investigation, as will be discussed for each polymer in particular in the discussion chapter. In general, two to three runs were performed for each analyzes using a sample weight of 5 to 10 mg. The determination of residual solvent amount in the coatings was performed by thermo gravimetric analyses using the NETZSCH TG 209 F1 equipment. The weight change was investigated in the temperature range of 25 - 500 °C at a heating rate of 10 K min^{-1} under argon atmosphere. The sample weight in all cases was in the range of 5 to 10 mg. For these thermal analyses the coatings were removed from the substrate using a sharp blade.

3.5 – Corrosion tests

3.5.1 - Electrochemical analysis.

The electrochemical corrosion behaviour of the samples was evaluated using a typical three-electrode cell as shown in Figure 3.3a. In this cell the sample was the working electrode (exposure area of 1.54 cm^2), a platinum mesh the counter electrode and a Ag/AgCl electrode was the reference one. The cell was connected to a potentiostat Gill AC from ACM instruments for the electrochemical measurements. When a high resistance polymer coating was investigated the cell was connected to a fempto amp device, which enhances the low current detection capacity, and placed inside a faraday cage to reduce noise in the spectra.

For regular analyses the corrosive solution was 3.5 wt.-% NaCl. Before the impedance test the open circuit potential (OCP) was measured for 15 - 30 minutes to let the potential stabilizes. Then the impedance test was carried out at amplitude of 10 mV for uncoated substrates and of 15 mV for coated substrates at frequencies ranging from 10^4 to 10^{-2} Hz.

To evaluate the performance of PAN coatings in biomedical applications, EIS tests were performed in simulated body fluid (SBF) with a given chemical composition. These analyses were performed using a special cell with a sample exposure area of 0.5 cm^2 and an external container that allow the flux of water to regulate the solution temperature. A thermostat was connected to this cell and the temperature was settled at 37.5 °C. The impedance test was carried out in the same way as described above. The impedance spectra of all the samples were simulated using the software Zview2 from Scribner associates to get more information on the corrosion mechanism. An image of the complete experimental setup is shown in Figure 3.3b.

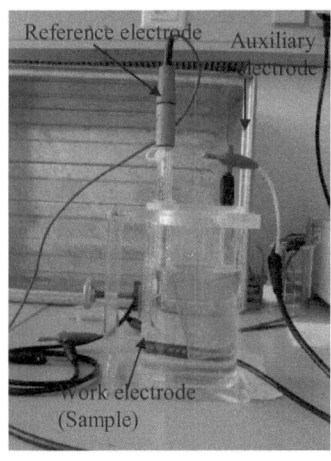

(a)

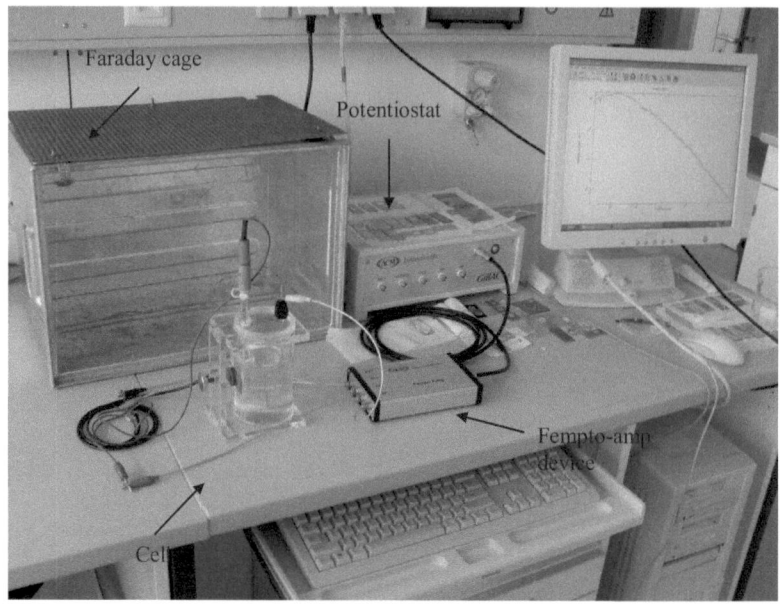

(b)

Figure 3.3 Image of the facility used for the electrochemical investigations.

For the uncoated substrates, 30 minutes after the impedance test was finished the polarization analyses were performed. A potential sweep was applied at a constant scan rate of 12 mV/min, starting 150 mV below OCP and finishing at the current limit of 0.1 mA cm^{-2}. Three to four measurements were carried out for each treatment condition as well as for the untreated sample.

3.5.2 – Immersion corrosion test

Besides the electrochemical corrosion tests, immersion tests were performed. These analyses consisted basically in completely immerse the sample in the corrosive solution and follow changes in the visual aspect of the samples with the time. This test has the advantage to allow the investigation of a whole sample instead of only specific areas. The investigations in 3.5% solution were performed at room temperature while the test with SBF was performed at 37 °C.

4 – Results

4.1 – Pre-treatments
4.1.1 - Hydrofluoric acid (HF) treatment
4.1.1.1 - Weight change and SEM analyses

The hydrofluoric acid treatment was selected due to its interesting properties as described in the previous chapter. As the optimal conditions for this treatment is not reported, different solution concentration and treatment times were tested. The immersion of Mg AZ31alloy in HF solutions resulted in a gas evolution at the starting period of the treatment. The gas emission is related to hydrogen formed by the reaction of HF with Mg, according to equation 4.1.1 [1.36, 1.42].

$$2HF + Mg \longrightarrow MgF_2 + H_2 \quad \text{equation 4.1.1.}$$

Figure 4.1.1.1a shows the samples weight change in relation to the treatment time, for different HF concentrations. After 1h of immersion, all solutions resulted in a weight loss, indicating a higher rate of material dissolution than of protective layer formation. After 5 h of immersion, a gradual weight gain (compared to the weight at 1h of treatment) was produced by the concentrations of 7 and 14 mol L^{-1}. This indicates that the rate of protective layer formation became higher than the rate of material dissolution. After 24 h of treatment, the samples treated with 7 and 14 mol L^{-1} have a weight gain of 0.14 mg cm^{-2} and 0.10 mg cm^{-2}, respectively, corresponding to a layer thickness of about 2 μm, as shown in Figure 4.1.1.1b.

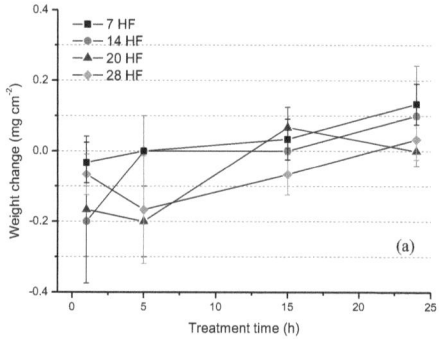

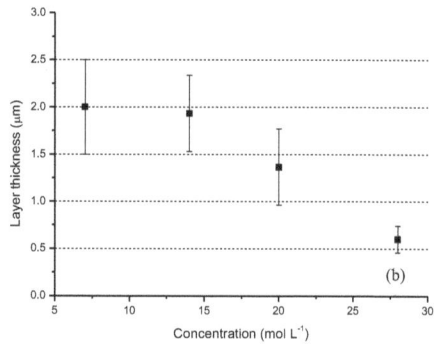

Figure 4.1.1.1: (a) Weight change of samples as a function of treatment time (the negative signal represents weight loss). (b) Layer thickness after 24 h of treatment as a function of solution concentration.

The samples treated with the concentrations of 20 and 28 mol L^{-1} HF showed weight increase only after 15 h of immersion. After 24 h, these solutions resulted in a weight gain of around 0.05 mg cm^{-2}. This smaller weight gain indicates that the rate of protective layer formation was lower for these concentrations compared to 7 and 14 mol L^{-1}. Figure 4.1.1.1b corroborates this result, showing that these concentrations produced thinner layers after 24 h of treatment. These results suggest that, for the used concentration range, the higher the HF concentration the lower the protective layer formation rate. This trend is probably associated to a higher material dissolution rate at higher acid concentrations.

Figure 4.1.1.1a also shows that the reaction of magnesium AZ31 alloy with HF has slow kinetics at the applied conditions. Only small quantities of protective layer (less than 0.2 mg cm^{-2}) were formed on the sample' surfaces, even after 24 h of treatment. Comparing to the work of Chiu et al. [1.42], who reported a weight gain of 35 mg cm^{-2} for pure magnesium ingots after 24 h of immersion in 28 mol L^{-1} HF, it can be concluded that the reaction of AZ31 with HF is much slower than that of pure magnesium ingots. This leads to the suggestion that the alloying elements Al, Zn and Mn improve the chemical stability of magnesium compounds in acidic fluoride environments, in a similar manner as they improve it in chloride environments, as reported by Pardo [1.22].

Figure 4.1.1.2 shows SEM images of a sample treated with 7 mol L^{-1} HF for different times. After 5 h of treatment it is possible to observe the deposition of compounds on the sample surface, especially compared to the surface after 1 h of treatment. After 15 h the entire surface was covered by a smooth layer, which got an irregular morphology after 24 h of treatment. This result is in agreement with the weight change measurements showing a gradual weight gain produced by the concentration of 7 mol L^{-1} after 5h of treatment. Similar analyses of samples treated with other HF concentrations also confirmed the weight change results.

The macroscopic aspect of the samples is shown in Figure 4.1.1.3. It can be observed that different acid concentrations results in different surface colours. The treatment with 7 and 14 mol L^{-1} HF resulted in a yellow collared surface while the treatment with 20 and 28 mol L^{-1} HF produced a black surface. This different aspect is an indicative of distinct compounds on the substrate surface and shows the importance of determining the appropriate treatment conditions, since different surfaces are created with the same acid at different concentrations.

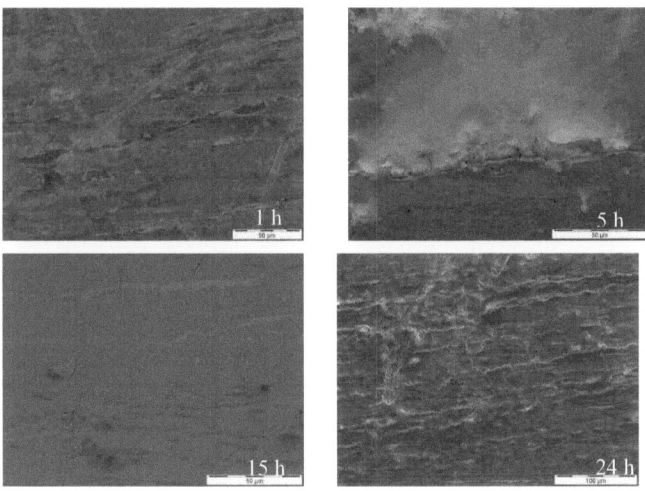

Figure 4.1.1.2: SEM figures of samples treated with 7 mol L^{-1} HF for different times.

Figure 4.1.1.3: Image of the samples treated with 14 mol L^{-1} HF (left) and 20 mol L^{-1} HF (right) for 24 h.

4.1.1.2 – OES analyses

The OES results of samples treated with 14 and 28 mol L^{-1} HF and of as-received and ground samples are given in Table 4.1.1.1. A slight decrease in magnesium concentration after the HF treatment can be observed, due to its dissolution. This magnesium dissolution resulted in an enrichment of some elements, like Al and Mn for the sample treated with 14 mol L^{-1}. As described in the introduction, iron, copper and nickel are the most deleterious impurities for magnesium alloys [1.17, 4.1, 4.2]. From Table 4.1.1.1 it can be seen that only the iron concentration was reduced by the HF treatment, reaching values close to that of the bulk composition (ground substrate), in particular, for the concentration of 28 mol L^{-1}. Nickel and copper were not dissolved during the treatment, probably due to their presence in more stable phases, as described by Liu et al. for copper [1.27]. It is also possible that the removal of these metals was not detected as their concentration is very low.

Table 4.1.1.1: Chemical composition of the samples, obtained by OES analyzes.

Sample	Mg (%)	Al (%)	Zn (%)	Mn (%)	Fe (%)	Ni (%)	Cu (%)
Untreated	95.70	3.23	0.823	0.225	0.008	0.001	0.001
Ground	95.72	3.23	0.810	0.222	0.005	0.001	0.001
14 HF	94.84	4.04	0.812	0.246	0.006	0.001	0.001
28 HF	95.31	3.62	0.829	0.214	0.005	0.001	0.001

In the introduction it is discussed that the Fe/Mn ratio is considered as a critical factor in corrosion studies of magnesium [4.3, 4.4]. Its critical value is reported as 0.032 for AZ alloys [4.5, 4.6]. In the present study, the treatment of magnesium AZ31 alloy with 14 and 28 mol L^{-1}

HF reduced the Fe/Mn ratio from 0.035 to 0.024 and 0.023, respectively, below the critical value, showing that this treatment is a cleaning process with respect to iron.

4.1.1.3 – FT-IR and XPS investigations

According to previous literature [1.42, 1.50], the treatment of Mg alloys with HF results in the formation of a MgF_2 layer at the metal surface. Magnesium fluoride has intense IR bands only below 600 cm^{-1}, but in the IR spectra of treated Mg AZ31 samples (Figure 4.1.1.4) there is a broad signal above 3000 cm^{-1}, a duplet at 2364 cm^{-1}, another signal at 1640 cm^{-1} and some broad signals below 900 cm^{-1}. This clearly indicates that not only MgF_2 was formed at the surface of the specimens.

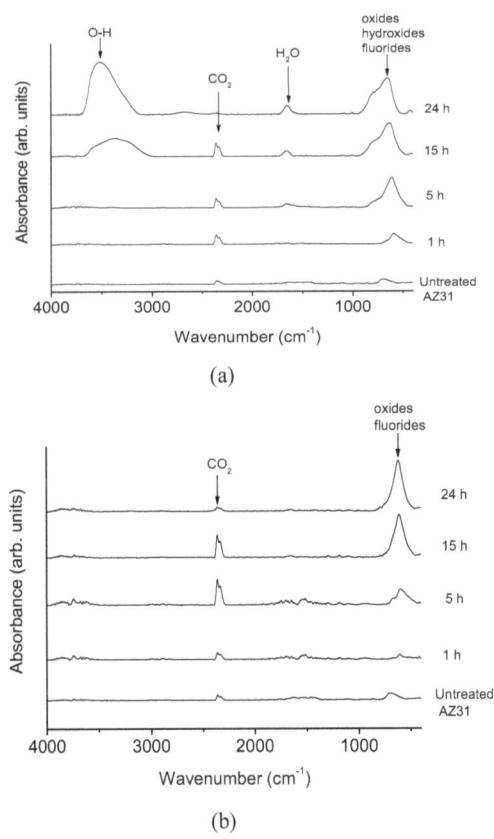

Figure 4.1.1.4: FT-IR spectra of Mg AZ31 samples treated with (a) 14 mol L^{-1} HF and (b) 28 mol L^{-1} for different times.

The broad signal above 3000 cm^{-1} is related to the O-H stretching of hydroxides and to water molecules linked to the surface by hydrogen bonds. The deconvolution of this signal, Figure 4.1.1.5a, results in four different O-H stretching modes. The signal at 3280 cm^{-1} can be attributed to adsorbed water, and the other three signals can be related to Mg(OH)$_2$, Al(OH)$_3$ and Zn(OH)$_2$ [4.7]. The intensity of these different O-H signals decreased with an increase of the HF concentration which is caused by the high hydrophobic character of the fluoride [4.8]. In general, an increase in the treatment time results in an increase in signal intensity. An exception is the concentrated HF (28 mol L^{-1}) where no hydroxides were observed (Figure 4.1.1.4b). The signal at 1640 cm^{-1} is related to the bend mode of the H-OH bond, which confirms the presence of adsorbed water. The presence of a shoulder at 1570 cm^{-1} indicates different interactions between water and hydroxides at the metal surface, which agrees with the high quantity of different O-H signals. The duplet at 2364 cm^{-1} is related to CO$_2$ adsorbed from the environment [4.9].

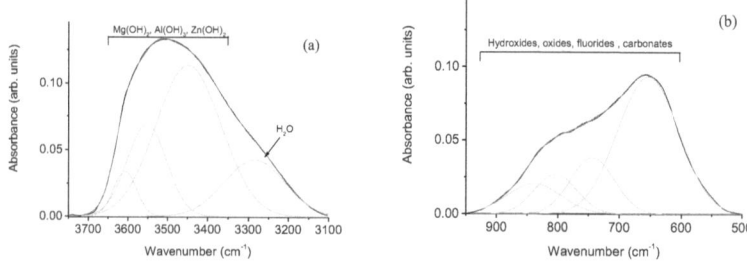

Figure 4.1.1.5: Deconvolution of the FT-IR spectrum of a sample treated with 14 mol L^{-1} HF for 24 h in the region of: (a) 3100 – 3700 cm^{-1} and (b) 500 – 950 cm^{-1}.

In general, below 900 cm^{-1} the spectra shows series of signals (Figure 4.1.1.5b), usually with an intense one at 650 cm^{-1} and three signals of lower intensity around 745, 800 and 840 cm^{-1}. By increasing the treatment time, these signals shifted to higher wave numbers. On the other hand, the signals shifted to lower wave numbers with increasing HF concentration. The number of signals in this range also increased with treatment time and decreased with acid concentration. This suggests that the signals that were present only at low HF concentrations and higher treatment times are related to hydroxides and/or to oxides.

According to previous literature, the signals below 900 cm^{-1} can be related to stretching and bending modes of hydroxide, oxide, carbonate and fluoride specimens [4.7, 4.9-4.12]. At this range, signals related to compounds of the general formula Mg(OH)$_x$F$_{2-x}$ (e.g. Mg(OH)$_{1.6}$F$_{0.4}$, Mg(OH)$_{1.2}$F$_{0.8}$) as described by Prescott *et. al* [4.13] is also reported. The presence of such compounds was investigated by XPS, and the results are shown in Table 4.1.1.2. Table 4.1.1.2 indicates that the F/Mg ratio varies from 1.7 to 2.0, depending on acid concentration and etching time (depth). As the F/Mg ratio of MgF$_2$ should be equal to 2, lower values indicate the presence of other magnesium compounds, e.g., Mg(OH)$_2$. As the O/Mg ratio varies from 0.2 to 0.3, in almost all samples, and the summation of O/Mg and F/Mg equals 2, considering a variation of 0.15 in the ratios, the presence of the compounds Mg(OH)$_{0.3}$F$_{1.7}$, Mg(OH)$_{0.2}$F$_{1.8}$ and Mg(OH)$_{0.1}$F$_{1.9}$ is suggested. Thus, it can be concluded that the formed layer is mainly constituted of magnesium fluoride, but hydroxide and oxides are present in the crystalline structure.

Table 4.1.1.2: F/Mg and O/Mg ratios calculated using the XPS results.

Etching time (s)	7 HF		14 HF		20 HF		28 HF	
	F/Mg	O/Mg	F/Mg	O/Mg	F/Mg	O/Mg	F/Mg	O/Mg
60	2.0	0.3	1.9	0.2	2.0	0.1	1.9	0.3
300	1.8	0.3	2.0	0.2	1.8	0.2	2.0	0.3
540	2.0	0.3	1.9	0.2	1.7	0.2	2.0	0.3
600	1.9	0.3	2.0	0.2	1.7	0.2	1.9	0.3

4.1.1.4 – Electrochemical investigations

Figure 4.1.1.6 shows the results of the impedance measurements of samples treated with 20 mol L^{-1} HF and 28 mol L^{-1} HF after 15 min of exposure to a 3.5 wt.-% NaCl solution. It can be observed that the sample treated with 20 mol L^{-1} HF for 24h has a much higher impedance compared to 28 mol L^{-1} HF and to the other treatment times. The samples treated with 14 mol L^{-1} HF behaved similar to those treated with 20 mol L^{-1} HF, and the higher impedance was obtained for these two conditions, as shown in Table 4.1.1.3, three orders of magnitude higher than that of the untreated sample. The samples treated with 7 mol L^{-1} HF showed a similar behaviour to the one observed in Figure 4.1.1.6b.

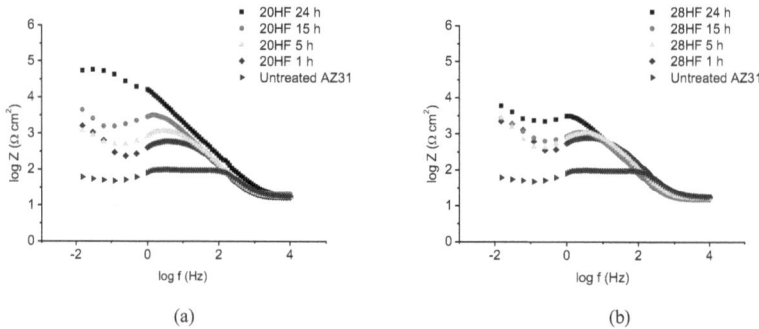

Figure 4.1.1.6: Bode plots of Mg AZ31 samples treated with 20 and 28 mol L^{-1} HF for different treatment times, after 30 minutes of exposure to a 3.5 wt.-% NaCl solution.

Figure 4.1.1.7 shows the impedance of samples treated with 14 and 20 mol L^{-1} HF during 24 h, at different exposure times to 3.5 wt.-% NaCl solution. After 20 h of exposure the impedance of the treated samples was similar to that of the untreated one, showing that the layer has already failed. This result shows that, the HF treatment is not suitable as a final corrosion protection process, because it does not result in long-term stable conversion coatings.

Additionally, the corrosion behaviour was evaluated by polarization measurements, as presented in Figure 4.1.1.8. In general, the polarization curves did not have a defined anodic slope and showed direct dissolution of the metal above the corrosion potential. This suggests the presence of defects in the protective layer. Table 4.1.1.3 shows the results of the cathodic slope analysis, where it can be seen that the samples treated with 20 mol L^{-1} HF and 14 mol L^{-1} HF for 24 h had the lower corrosion current, 0.017 and 0.019 mA/cm^2, respectively, 6 times lower than that of the untreated sample, corroborating the results of the impedance spectra.

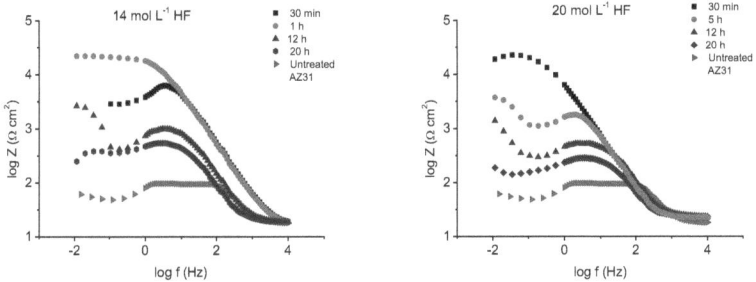

Figure 4.1.1.7: Bode plots of Mg AZ31 samples treated with 14 and 20 mol L^{-1} HF at different exposure times to a 3.5 wt.-% NaCl solution.

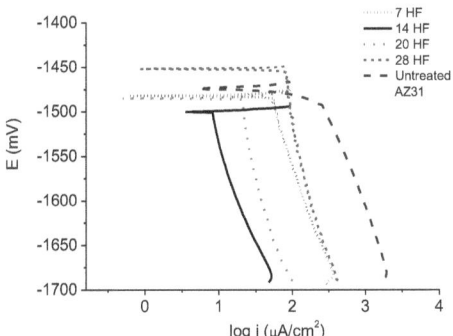

Figure 4.1.1.8: Polarization curves of the samples treated with HF during 24 h.

For all samples, no correlation between the corrosion potential, E_{corr}, and the corrosion current was observable. Considering these two "best" acid concentrations and taking into account safety and economic aspects, the 14 mol L^{-1} HF is preferable than 20 mol L^{-1} HF since it is less aggressive and demands less hydrofluoric acid, which is relative expensive (800 mL costs approx. 200 Euros). Therefore, this solution was selected for the pre-treatment of the samples before the coating process.

To evaluate if a prior cleaning could enhance the protective properties of this treatment, the 20 mol L^{-1} HF solution was also used to treat ground samples. The results of the electrochemical tests are given in Table 4.1.1.3. By comparing these results with those of the as-received samples it can be observed that the grinding process had no beneficial effect on the corrosion resistance. Instead, a higher corrosion rate was obtained. This is related to the removal of the partially protective MgO film during the grinding process. During the HF treatment, besides impurities removal (in the first hours of immersion), a part of the MgO film present on the as-received sample is maintained. The grinding removes impurities (as does the HF treatment) but completely removes the partially protective MgO film. For this reason, the ground samples showed slightly worse behaviour. This is an interesting result as it demonstrates that the HF treatment can be applied directly on the as-received material without requiring prior cleaning.

Table 4.1.1.3: Electrochemical results obtained by the impedance spectra and polarization curves. In the Table, "Z" corresponds to the impedance at the lowest frequency on the ESI spectra.

Concentration	Z (kΩ cm^2)	E$_{corr}$(mV)	I$_{corr}$(mA/cm^2)
7 HF 1 h	0.56	-1510 ± 20	0.074 ± 0.029
7 HF 5 h	0.93 ± 0.33	-1525 ± 14	0.049 ± 0.023
7 HF 15 h	4.00 ± 1.5	-1523 ± 16	0.033 ± 0.023
7 HF 24 h	1.88 ± 0.41	-1470 ± 16	0.055± 0.004
14 HF 1 h	0.547 ± 0.262	-1520 ± 11	0.039 ± 0.029
14 HF 5 h	2.11 ± 1.60	-1535 ± 20	0.023 ± 0.010
14 HF 15 h	3.88 ± 0.454	-1507 ± 6	0.030 ± 0.014
14 HF 24 h	85 ± 50	-1498 ± 23	0.019 ± 0.01
20 HF 1 h	0.58 ± 0.04	-1513 ± 1.7	0.029 ± 0.001
20 HF 5 h	1.39 ± 0.37	-1506 ± 6	0.033 ± 0.001
20 HF 15 h	3.42 ± 1.68	-1468 ± 23	0.013 ± 0.011
20 HF 24 h	65 ± 11	-1468 ± 28	0.017 ± 0.005
28 HF 1 h	0.90 ± 0.54	-1516 ± 16	0.025 ± 0.012
28 HF 5 h	1.14 ± 0.01	-1525 ± 44	0.029 ± 0.003
28 HF 15 h	1.11 ± 0.13	-1445 ± 22	0.030 ± 0.009
28 HF 24 h	2.92 ± 1.30	-1459 ± 32	0.062 ± 0.034
Untreated AZ31	0.06 ± 0.01	-1473 ± 1	0.111 ± 0.010
20HF 24 h (ground)	75 ± 2.01	1449 ± 69	0.020 ± 0.004

4.1.2 – Grinding and acid cleaning

Aiming to compare the HF-treatment with other acid cleaning processes, acetic and nitric acid were used for the pre-treatment of the alloy. Theses acids were selected due to their interesting properties as described by Nwaogu [1.33, 1.34]. A detailed study on the influence of different concentrations and treatment times of AZ31 alloy using these two acids is reported by Nwaogu et al.[1.33, 1.34]. It was shown that acetic acid, in a concentration of 5 mol L^{-1} results in the most significant decrease in corrosion rate among the organic acids. In the group of inorganic acids, the lower corrosion rate was obtained using nitric acid at a concentration of 1 mol L^{-1}. For both acids the optimal treatment time was determined as 2 minutes. A complete characterization of the alloys treated with these acids is not presented in the present study since that this is done in the studies of Nwaogu et al. [1.33, 1.34].

From figure 4.1.2.1 it is obvious that both acids reduce the cathodic current compared to the as-received substrate. Despite similar values, nitric acid reduces the cathodic current slightly more than acetic acid. Nevertheless, both acids produced samples with substantial higher cathodic currents compared to the ground substrate and to the 14 mol L^{-1} HF treated sample. Thus, the best substrate treatments, with regard to low corrosion current, are classified in the following order: 14 mol L^{-1} HF > grinding > HNO_3 ≥ Acetic acid > as-received. Similarly to the results using different HF concentrations, no correlation between cathodic current and the corrosion potential could be evaluated.

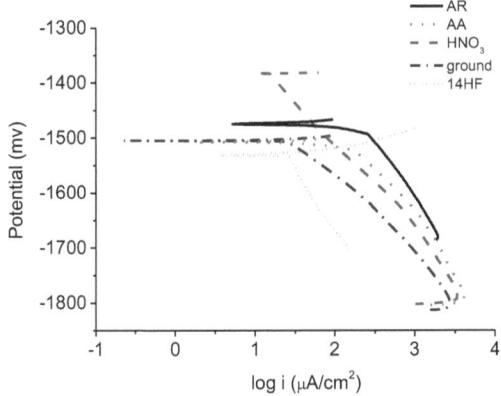

Figure 4.1.2.1: Polarization curves of ground and as-received (AR) substrates and of samples treated with acetic acid (AA), nitric acid and 14HF.

In table 4.1.2.1 it is shown the influence of these treatments on the substrate surface roughness. It can be observed that the treatment with acetic acid produced the higher surface roughness increase while the grinding process produced the higher surface roughness decrease. The treatment with HF and nitric acid did not result in considerable surface roughness variations, compared to the as-received sample. This difference in surface roughness is related to the metal dissolution rate in the acid solutions. The lower dissolution rate in HF is caused by the formation of the stable MgF_2 layer, while in HNO_3 it is related to the lower acid concentration in comparison to the acetic acid solution. In case of acetic acid, the 5 mol L^{-1} solution produced many bubbles and heat even in the first seconds of treatment, while the other acid treatments were much milder. The differences in surface roughness are important to evaluate the influence of this parameter in the coating adhesion and also on the coating morphology as will be discussed in details later on.

Table 4.1.2.1: Substrates surface roughness.

Sample	Surface roughness (μm)
As-received AZ31	0.37 ± 0.10
Ground AZ31	0.09 ± 0.01
Nitric Acid	0.36 ± 0.07
Acetic acid treated AZ31	2.12 ± 0.23
14HF treated AZ31	0.37 ± 0.02

Figure 4.1.2.2 gives the infrared spectra of the as-received, acetic and nitric acid treated substrates. It is visible that, while acetic acid considerably decreases the intensity of the signals below 1000 cm^{-1}, the treatment with nitric acid results in an intensity increase of these signals, and signals related to the NO_3^- anion appears. This indicates that acetic acid removes the partially protective magnesium oxide layer on the substrate surface while the treatment with nitric acid forms these compounds. It can also be observed that both acids increase the intensity of the signals higher than 3000 cm^{-1} indicating an increase in the concentration of hydroxides. In general, it can be said that the treatment with nitric acid forms more compounds on the substrate surface than acetic acid, which has a much stronger effect in the removal of the native magnesium oxide/hydroxide film. Similar conclusion were made by Nwaogu et al, who demonstrate that the removal of material was much higher in case of acetic acid compared to nitric acid [1.33, 1.34].

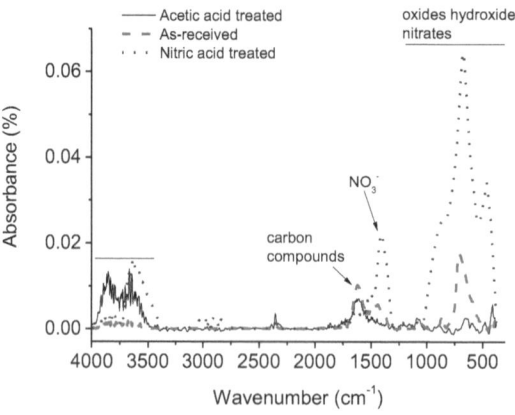

Figure 4.1.2.2: Infrared spectra of the as-received, acetic acid and nitric acid cleaned substrates.

4.2 – Polymer coatings
4.2.1 – Spin-coated poly (ether imide) [PEI]
4.2.1.1 – Coating characterization

The spin-coating process is a useful and common tool for the preparation of extreme thin and uniform films on flat substrates. This coating method is used in electronics industry, e.g. wafer coating, and for the preparation of solid oxide fuel cells (SOFC) [4.14-4.17]. One of the main advantages of this method is the thickness uniformity of the prepared coating, which is a problem for other coatings methods, as dip-coating, where the solvent flow induces non uniform covering of the surface, and consequently defect formation [1.63]. Besides that, the thickness of the coatings can be controlled by spinning speed and solution concentration, what allows the formation of uniform coatings with specific thicknesses. As a drawback, this method is restricted to flat substrates and it is not suited for large production rates since it is a batch process [1.63].

The morphology of PEI coatings prepared by spin-coating is extremely influenced by atmospheric humidity, as can be seen in figure 4.2.1.1. Under standard room conditions with certain humidity, the coatings have a white appearance indicating a porous morphology, as could be confirmed by the SEM images shown in figures 4.2.1.2a and 4.2.1.2c. These pores, with ca 2 µm of diameter, are formed due to the polymer precipitation in presence of air humidity [1.68, 4.18]. A similar behaviour is reported by Eisenbraun [1.75, 1.74] for polyamic acid and fluorinated polyimide. Under N_2 atmosphere, the dry gas induces a phase inversion process governed by solvent evaporation that leads to a transparent, non-porous coating, as can be observed in figures 4.2.1.2b and 4.2.1.2d.

Figure 4.2.1.1: A AZ31 sample spin-coated with PEI/NMP (10/90) under room atmosphere (left) and under N_2 atmosphere (right).

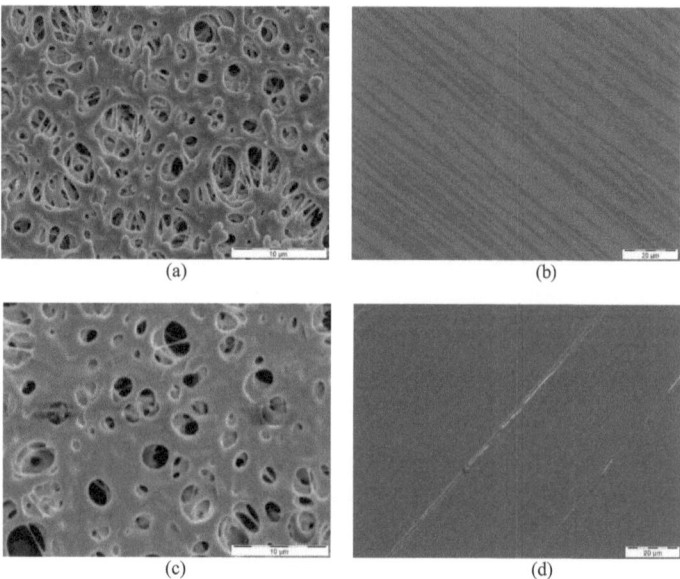

Figure 4.2.1.2: SEM photographs of spin-coated samples prepared on ground substrates. (a) and (b) are samples coated with PEI/DMAc (10/90) at room and N_2 atmosphere, respectively, and (c) and (d) are samples coated with PEI/NMP (10/90) at room and N_2 atmosphere, respectively.

The thickness of the coatings varies with the spinning speed, atmosphere, solution concentration and solvent type as shown in figure 4.2.1.3. For all the conditions, the thickness decreases with the spinning speed, due to the increase in the centrifugal force acting on the solution. A higher centrifugal force results in a higher solution outflow from the substrate, thinning the film. On the other hand, a higher spinning speed increases the solvent evaporation rate, and consequently, the solution viscosity. The viscous drag acts in the opposite direction of the centrifugal force inhibiting and excessive thinning of the coating. The influence of the atmosphere can be observed by comparing figures 4.2.1.3a with 4.2.1.3b. The higher thickness obtained at room atmosphere is related to the faster rate of polymer precipitation, induced by air humidity, than of solvent evaporation under N_2 atmosphere. After the polymer precipitation, the film could not undergo a further thinning process, and hence, the thickness at this condition was higher. The effect of humidity on the polymer precipitation was stronger for PEI/DMAc (10/90) than for PEI/NMP (10/90) due to the lower stability of the PEI/DMAc solution in the presence of water, as reported by Wang [4.19]. For this reason this solution resulted in thicker coatings. Solutions of 15 wt.-% were not used at room conditions due to non-uniform covering of the substrate, as observed in previous tests. This is probably related to a very high polymer precipitation rate of this solution at room atmosphere which inhibited the formation of a uniform layer.

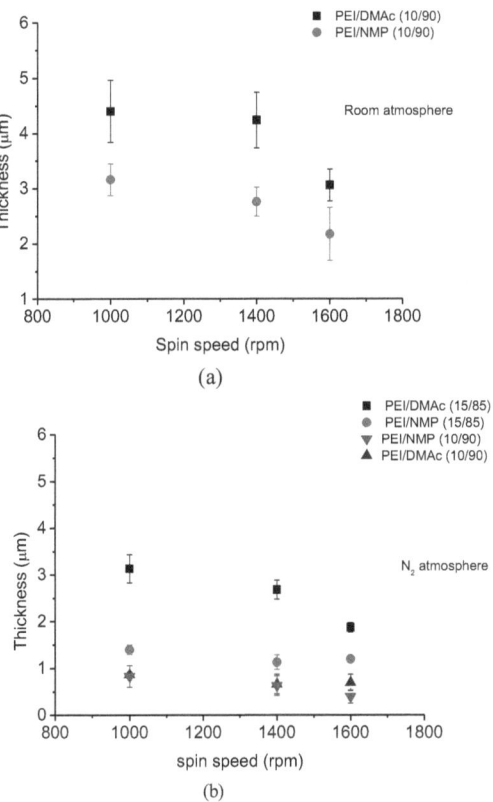

Figure 4.2.1.3: Thickness of the coatings as a function of spinning speed: (a) Room atmosphere and (b) N_2 atmosphere.

The influence of the solution concentration on the coating thickness, observed for the coatings prepared under N_2 atmosphere, figure 4.2.1.3b, is associated with the solution viscosity. According to studies in the literature, for the same solute-solvent system and the same spinning speed, the higher the viscosity the thicker the film [1.70, 1.71]. This is due to the lower outflow of a more viscous solution. As expected the 15 wt.-% solutions had higher viscosities than the 10% wt solutions (Table 4.2.1.1), and for that reason, produced thicker coatings.

Despite the lower initial viscosity of the solution PEI/DMAc (15/85) compared to PEI/NMP (15/85 (table 4.2.1.1), the first one produced thicker coatings. This result is related to the extra drying process applied on the samples coated using NMP solutions which produces a further thinning in these coatings. Besides that, the solvent evaporation during the spin-coating process leads to a viscosity increase [4.20]. As shown in table 4.2.1.1, DMAc has a much lower boiling point than NMP, being more volatile. Hence, at the beginning of the spin-coating process, PEI/DMAc (15/85) undergoes a higher viscosity increase than PEI/NMP (15/85), due to its higher solvent evaporation rate. This viscosity increase also collaborates for the higher thickness of the coatings prepared using DMAc [1.78].

Table 4.2.1.1 Solutions used in the coating process.

Polymer/Solvent	Concentration (wt.-%)	Viscosity (Pa s)
PEI/DMAc	10	0.043
	15	0.245
	20*	1.503
PEI/NMP	10	0.084
	15	0.620
	20*	2.980

* Used only for dip-coating

Figure 4.2.1.4 shows the FT-IR spectra of coatings prepared using NMP and DMAc as solvents, before and after drying in a vacuum oven at 135°C during 12 h. In figure 4.2.1.4a it is show the influence of the solvent on the signal around 1360 cm^{-1} which is related to the C-N-C stretching mode of the imide ring [4.21]. This signal usually appears centred at 1365 cm^{-1} with a small shoulder around 1380 cm^{-1}, as it appears in figure 4.2.1.4b. For the coating prepared using NMP this signal splits into three ones. This suggests that NMP strongly interacts with the imide ring of PEI that makes it difficult to remove it by spin-coating and even by the drying in the vacuum oven, as the signal around 1360 cm^{-1} still indicates residual amounts of NMP. Such interaction is not observed for the coating prepared using DMAc. Thermo gravimetric analyses showed a slightly higher amount of residual NMP than DMAc in the respective coatings prepared using 15 wt.-% solutions (6.07% and 5.49% respectively).

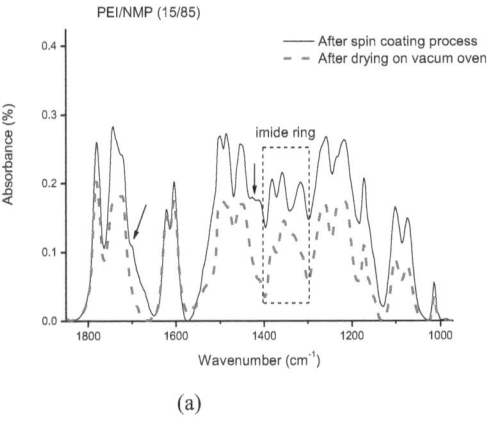

(a)

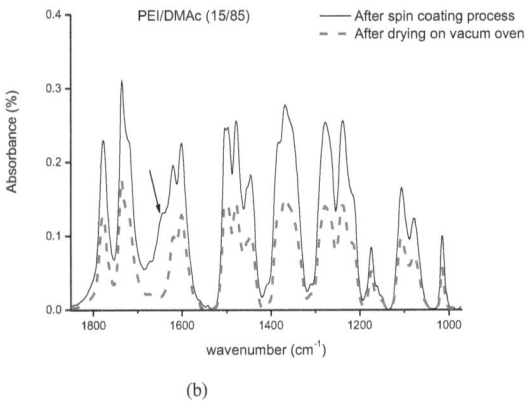

(b)

Figure 4.2.1.4: FT-IR spectra of spin-coated samples (ground substrates) before and after drying in a vacuum oven at 135 °C for 12h. The arrows in the figures indicate signals related to the solvents.

4.2.1.2 – Electrochemical impedance spectroscopy (EIS)

Figure 4.2.1.5 shows the impedance spectra of samples prepared by spin coating under room and N_2 atmosphere. The impedance of the samples coated with PEI/DMAc (10/90) at room conditions, figure 4.2.1.5a, is in the order of 10^5 Ω cm^2, in the same magnitude of PEO coatings described in the literature [1.53-1.57], and only one order of magnitude lower than that of coatings prepared under N_2 atmosphere (figure 4.2.1.5b). Nevertheless, despite this considerably high initial impedance, the coatings prepared at room atmosphere did not show acceptable protective properties after one day of exposure to the corrosive solution, and the coatings morphology was highly sensitive to humidity variations, affecting the reproducibility of the results. For this reason, further characterization was focused on coatings prepared under N_2 atmosphere. As figures 4.2.1.5a and b show the initial impedance does not considerably change with the spinning speed (coating thickness) a Mean value was selected for further studies (1400 rpm).

Due to their non-porous morphology the samples prepared under N_2 atmosphere showed good corrosion protection after 2-9 days of exposure to the corrosive solution, depending on the coating thickness. Figure 4.2.1.6 shows the impedance variation with time of coatings prepared using DMAc as solvent. The decrease in impedance with exposure time indicates the diffusion of water and ions through the coating which increased its dielectric constant, and consequently, decreased its resistance [1.95]. After 20 h of exposure to the corrosive solution, the samples spin-coated with PEI/DMAc (10/90), figure 4.2.1.6a, showed impedances in the order of 10^4 Ω cm^2 and a capacitive behaviour in a small frequency range (10^3 - 10^4 Hz) indicating that the coating considerably had lost its capacity to separate the solution from the metallic substrate [1.67, 1.95]. After 48 h of exposure, this coating was completely damaged.

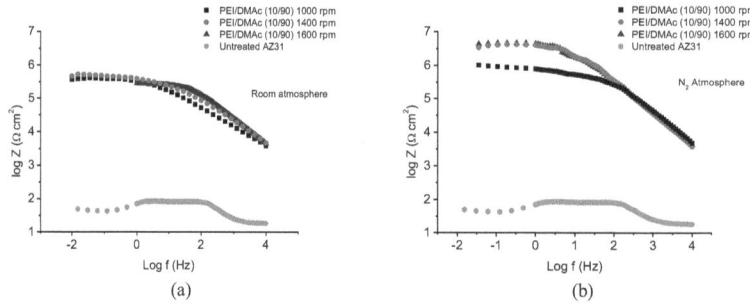

Figure 4.2.1.5: Impedance spectra of samples spin-coated with PEI/DMAc (10/90) on ground substrates, at different atmospheres and spinning speeds. The measurements were performed after 15min of exposure to a 3.5 wt.-% NaCl solution.

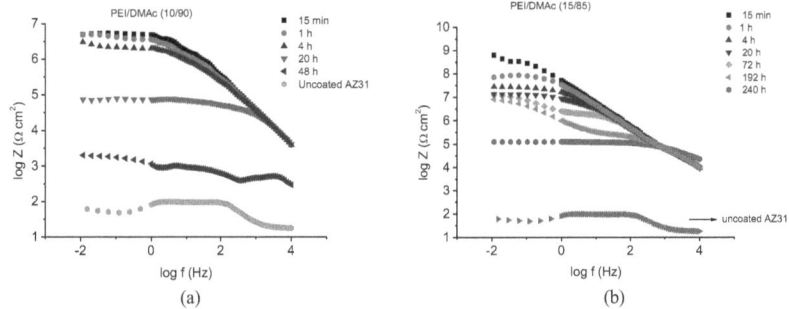

Figure 4.2.1.6: Impedance spectra of samples spin-coated at 1400 rpm on ground substrates using (a) PEI/DMAc (10/90) and (b) PEI/DMAc (15/85), under N_2 atmosphere at different exposure times to a 3.5% NaCl solution.

The samples coated using PEI/DMAc (15/85), figure 4.2.1.6b, showed better result due to their increased thickness. The initial impedance was in the order of 10^9 Ω cm^2, which is in the same order of PEI coatings prepared by dip-coating using CH_2Cl_2 as solvent, reported in a previous publication of our group [1.68]. At 72 h of exposure the spectra started to change at frequencies below 10^0 Hz (of logf = 0), showing another capacitive part (-1 slope), indicating the gradual concentration of water and anions on the metal/polymer interface. At this exposure time, the impedance dropped to 10^7 Ω cm^2 and maintained this value for the next 192 h, which is an excellent result for coated magnesium compounds compared to other reports in the literature [1.53-1.57, 1.68] especially at this low coating thickness. The impedance reaches 10^5 Ω cm^2 after 240 h of exposure and the value of the uncoated metal 24 h after that. The performance of this sample was confirmed by measuring two other samples which showed similar behaviours.

The samples coated with PEI/NMP (10/90), figure 4.2.1.7a, showed impedance values of 10^4 Ω cm^2 after only 4 h of exposure, indicating that the solution could easily penetrate and reach the substrate. This was confirmed by a second capacitive behaviour that appeared after 1h of exposure in the frequency range of 10^0–10^2 Hz. After 10 h of exposure, the coatings prepared using PEI/NMP (10/90) showed considerable degradation. The samples coated by PEI/NMP (15/85) showed inferior behaviour compared to PEI/DMAc (15/85). They were considerable degraded after 48 h of exposure (figure 4.2.1.7b). The reason for the inferior behaviour of the coatings prepared using NMP in both concentrations is related to residual solvent and, in the case of coatings prepared using 15%-wt solution, to their lower thickness.

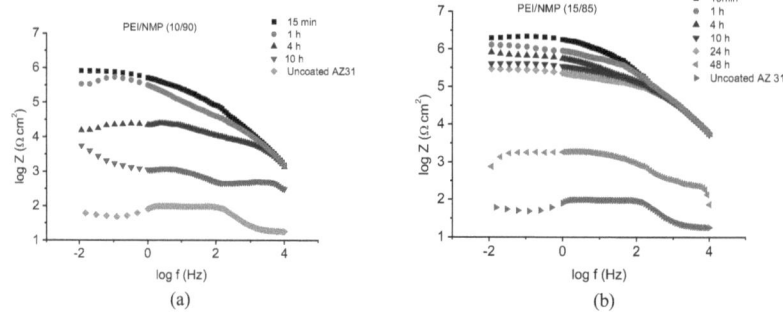

Figure 4.2.1.7: Impedance spectra of samples spin-coated at 1400 rpm on ground substrates using (a) PEI/NMP (10/90) and (b) PEI/NMP (15/85) under N_2 atmosphere at different exposure times to a 3.5% NaCl solution.

Despite the good performance of the coating prepared using DMAc, a much better performance was obtained when the coating was post-dried in a vacuum oven, as can be observed in figure 4.2.1.8 (the residual solvent amount was reduced from 6% to 4% by this process). This coating maintained impedance in the order of 10^6 cm^2 even after 768 h of exposure to the corrosive solution. Similar impedances are only described in the literature for much thicker and more complex coatings. Due to the coating transparency, it was possible to observe that corrosion products start to be formed on the substrate at 336 h of exposure. However, the presence of these corrosion products has significant influence on the coating deterioration only at an exposure time of 768 h when the coating starts to show considerable impedance decrease in a time period of 48 h. The better performance of the post-dried coating shows that the presence of residual solvent plays a significant role in the protective properties of the coatings. A total removal of the residual solvent could be achieved using higher temperature and longer drying times. This would certainly reduce even more the diffusion of water but could have negative influences in the mechanical properties of the film (e.g. increasing brittleness).

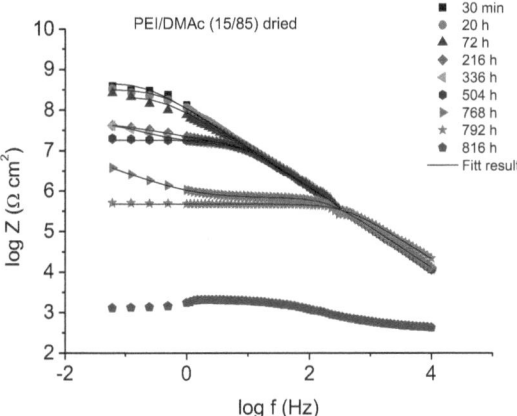

Figure 4.2.1.8: EIS spectra of a coating prepared using PEI/DMAc (15/85) after 2 h in a vacuum oven at 135 °C.

To extend the understanding on how the presence of residual solvent affects the coating performance, EIS simulations of the just spin-coated and post dried samples were performed. In figure 4.2.1.9a it can be observed the theta Bode plot of the just spin-coated sample and the fitting result obtained using the circuit model shown in figure 4.2.1.9b. This circuit differs from the one used for the simulation of post-dried coating by an extra constant phase element (CPE) in parallel with other resistance, which are here called of CPE_1 and R_1, respectively. The addition of CPE_1 and R_1 in parallel was necessary to correctly simulate the high frequency behaviour observed in figure 4.2.1.9a. At high frequencies, theta shows two distinct loops which are observed in all immersion times (240 h is the only exception). At certain exposures time, the theta Bode plot clearly shows three time constants that were impossible to simulate using the traditional two CPEs circuit (Figure 1.11b). As shown in figure 4.2.1.10 these two high frequency capacitive loops are not present in post-dried samples, indicating that CPE_1 and R_1 simulates the influence of solvent rich domains in the coating capacitance.

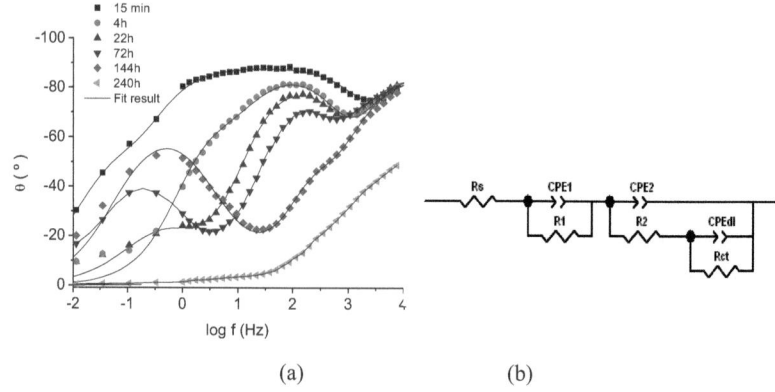

Figure 4.2.1.9: (a) Theta Bode plot of the samples coating with PEI/DMAc (15/85), (b) electronic circuit used in the results simulation.

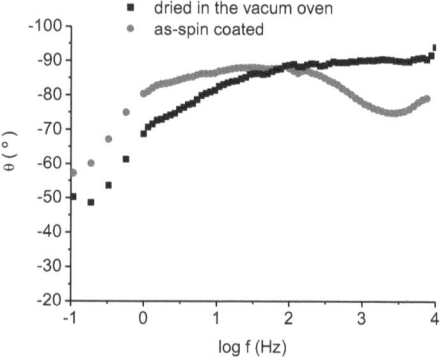

Figure 4.2.1.10: Theta Bode plot of just spin-coated and dried in a vacuum oven at 135 °C for 2h.

The fitting values of the just spin-coated and post-dried samples prepared using PEI/DMAc (15/85), are shown in table 4.2.1.2. It can be observed that, for the just spin-coated sample, T_1 maintains a constant value in the first 3h of exposure, show a slightly increase after 72 h and decreases after 144 h while T_2 shows a progressive increase during the whole exposure time. This constant value of T_1 in the beginning of exposure suggests that the solvent-rich domains on the coating surface are saturated with electrolytes. At this saturation stage, when an electrolyte enters in the solvent-rich domain of the coating an equal amount goes from it to the solution and/or to the solvent-poor domain inducing an increase in T_2 and a constant value of T_1. Considering that the solvent-rich domains on the coating surface are saturated, the following variations in T_1 are entirely related to changes in the A/d ratio of equation 1.4. The increase in T_1 observed until 72 h suggests that the electrolytes diffuses in the film and reaches others solvent-rich domains, increasing the area of saturated solvent-rich domains, and consequently, producing an increase in T_1. This is schematically shown in figure 4.2.1.11.

After certain exposure time the solvent-rich domains will be saturated in the entire coating exposed area (case 2 in figure 4.2.1.11) and the following variations in T_1 will be entirely related to changes in thickness. As the electrolytes get deep into the coatings, the total thickness of the saturated solvent-rich domains (the total thickness of all the saturated solvent-rich domains) increases, leading to the decrease in T_1 that is observed at 144 h. On the other hand, the value of T_2 increases constantly during the whole exposure time due to the increase in dielectric constant produced by the flux of electrolytes from the solution and saturated solvent-rich domain to the solvent-poor domain.

This model predicts that after certain exposure time, all the solvent-rich domains will be saturated with electrolytes in the whole volume of the exposed coating and T_1 will be constant. At this time, electrolytes will diffuse from the solvent-rich domain to the solvent-poor one until they have the same electrolyte amount, and CPE_1 and CPE_2 will merge in one single CPE. This is confirmed by the curve for 240 h of exposure shown in Figure 4.2.1.9a where only one time constant can be observed at high frequencies, indicating an equal electrolyte distribution in the coating. This curve was simulated using the tradition two CPE circuit (figure 1.11b) and it can be observed from Figure 4.2.1.10a and Table 4.2.1.2 that the model fitted very well in the curve and that the values follow the trend observed in the other exposure times.

Table 4.2.1.2: Results of the fitting process of the EIS spectra. In the table, T_1, T_2 and T_{dl} are the T coatings of CPE_1, CPE_2 and CPE_{dl}, respectively.

Exposure time	T_1 (Ω^{-1} cm^2)	R_1 (Ω cm^2)	T_2 (Ω^{-1} cm^2)	R_2 (Ω cm^2)	T_{dl} (Ω cm^2)	R_{dl} (Ω cm^2)
Just spin-coated						
15 min	4.0 x 10^{-9}	1.2 x 10^4	3.0 x 10^{-9}	3.0 x 10^8	5.0 x 10^{-9}	1.0 x 10^9
3 h	4.0 x 10^{-9}	3.0 x 10^4	4.0 x 10^{-9}	6.5 x 10^6	8.0 x 10^{-9}	2.0 x 10^7
22 h	4.5 x 10^{-9}	3.0 x 10^4	4.6 x 10^{-9}	4.0 x 10^6	1.2 x 10^{-7}	6.9 x 10^6
72 h	6.6 x 10^{-9}	5.0 x 10^4	4.8 x 10^{-9}	1.8 x 10^6	1.8 x 10^{-7}	1.2 x 10^7
144 h	4.3 x 10^{-9}	6.0 x 10^4	8.0 x 10^{-9}	2.4 x 10^5	2.5 x 10^{-7}	8.5 x 10^6
240	-	-	2.2 x 10^{-8}	1.2 x 10^5	9.2 x 10^{-6}	4000
Post-dried						
30 min	-	-	1.6 x 10^{-9}	1.2 x 10^9	-	-
20 h	-	-	1.4 x 10^{-9}	2.9 x 10^8	-	-
72 h	-	-	1.6 x 10^{-9}	1.2 x 10^8	-	-
216 h	-	-	1.2 x 10^{-9}	1.5 x 10^7	2.8 x 10^{-8}	3.6 x 10^7
336 h	-	-	1.3 x 10^{-9}	1.5 x 10^7	3.9 x 10^{-8}	4.2 x 10^7
504 h	-	-	1.3 x 10^{-9}	1.7 x 10^7	4.5 x 10^{-7}	5.6 x 10^6
768 h	-	-	1.2 x 10^{-9}	6.7 x 10^5	5.1 x 10^{-7}	1.2 x 10^7
792 h	-	-	1.37 x 10^{-9}	4.69 x 10^5	-	-

Interesting considerations can also be drawn regarding R_1 and R_2. Table 4.2.1.2 shows that R_1 increase while R_2 decreases during all exposure times. The decrease of R_2 is a normal and expected behaviour related to the increase in the electrolyte content in the solvent-poor domain. In the case of R_1, the increase is probably related to solvent been washed out from the coating. When the solvent is washed out, the resistance of the solvent-rich domain moves towards the resistance of the solvent-poor one. However, as the increase was small compared to R_2, it can be conclude that only a small amount of solvent was washed out until 144 h of exposure. It is interesting to observe that R_2 is three orders of magnitude higher than R_1 at the first hours of exposure, but after 144 h, they are in a similar range. This is another indicative that the electrolytic distribution becomes more homogeneous with the exposure time, and that the differences in electrolytes contents between solvent-rich and solvent-poor domains becomes smaller until only one CPE is observed.

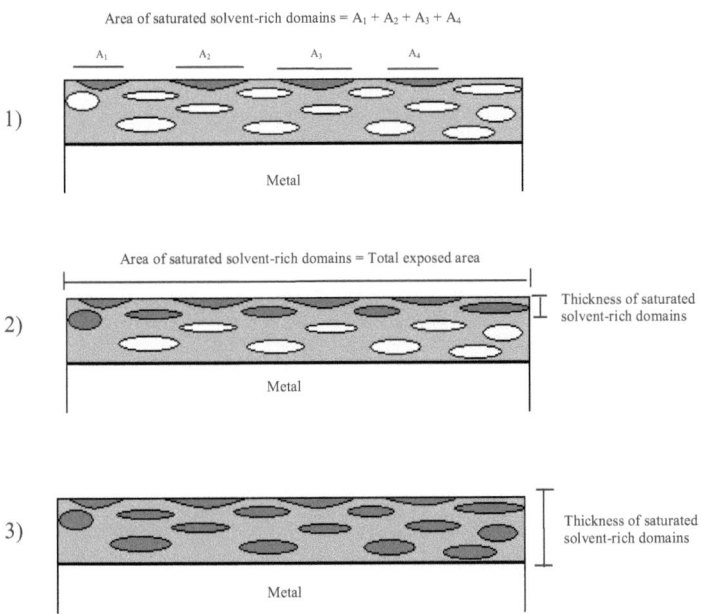

■ Saturated solvent-rich domains; ☐ Solvent-rich domains

Figure 4.2.1.11: Scheme of the electrolyte diffusion in the coating: 1) the electrolyte starts entering the coating saturating the solvent-rich domains at the surface; 2) the electrolyte diffuses in the coating saturating other solvent rich domains right beneath the surface; 3) the electrolyte moves deep into the coating, saturating all the solvent-rich domains in the whole coating volume exposed to the solution.

Beside these two coatings constant phase elements, a third one related to the double-layer of ions in the polymer-metal interface is observed. This double layer CPE appears even after 15 min of exposure, suggesting the presence of small defects in the coating which allowed the fast arrival of electrolytes in the interface. As expected, T_{dl} increases and R_{ct} decreases with exposure time, indicating the increase of electrolyte at the interface.

The behaviour of samples coated by PEI/NMP (15/85) follow a similar trend as the one described above, but however, a much higher variation in T_1 and T_2 is observed in a shorter period of time, indicating a faster diffusion of water. This is related to a stronger interaction between PEI and NMP [4.19] that probably leads to a stronger plasticizing effect. This higher water diffusion rate is responsible for the low protective properties of this coating.

This could be further confirmed by comparing the wet adhesion values for these coatings, shown in table 4.2.1.3. As the adhesion is similar for both coatings it can be concluded that the rate determining effect of the impedance decreases with exposure time is not adhesion loss, but rather the difference in electrolytes diffusion in the coating. The higher electrolyte diffusion through the coatings prepared using NMP leads to higher currents and consequently lower impedance. The performance of PEI coatings prepared using NMP solutions will be certainly better using a drying step in a vacuum oven, but however, due to the lower thickness it is expected that the performance of the post-dried NMP coating will be worse than that of the post-dried DMAc one. Besides that, the NMP coatings would still require an extra drying step after spin-coating before the oven drying, otherwise the coating would be wet and could form a porous morphology by phase inversion caused from the contact with air and humidity. As a result, DMAc solution is much more appropriate for the preparation of PEI protective coatings.

Table 4.2.1.3: Adhesion test results.

Sample	Adhesion strength (MPa)
PEI/DMAc (15/85)	2.00 ± 0.64
PEI/NMP (15/85)	2.25 ± 0.51

In case of the post dried coating, the T_2 constant does not follow a regular variation and shows much lower values than in the just spin-coated sample. This irregular variation of T_2 suggests a non-uniform diffusion of electrolyte in the coating while the lower value is a result of the lower residual solvent amount. A detailed description on the degradation mechanism of the post dried coating will be given in the following chapters, since that it is the same as the degradation mechanism observed for PEI coatings prepared by the dip-coating method.

4.2.1.3- Influence of substrate pre-treatment

Figure 4.2.1.12 shows the pre-treatment effect on the coating morphology. The pre-treatment is an important process for the corrosion protection of coatings, since it influences the adhesion, the film formation process and the impurity concentration on the metal surface. It can be observed that the coatings prepared on acid treated substrates have a non uniform morphology due to surface roughness effects. The acetic acid treatment resulted in a higher average surface roughness (2.21 μm) compared to HF (0.37 μm) and HNO₃ (0.36 μm) inducing a higher coating surface roughness, as can be observed in figure 4.2.1.12a and table 4.2.1.4.

Table 4.2.1.4: Surface roughness.

Substrate	Uncoated substrate roughness* (μm)	Coated substrate Roughness* (μm)
Ground	0.09 ± 0.01	0.03 ± 0.01
14HF-Treated	0.37 ± 0.02	0.26 ± 0.09
HNO₃	0.36 ± 0.07	0.28 ± 0.01
Acetic acid cleaned	2.21 ± 0.23	0.97 ± 0.09

*The roughness values corresponds to the average surface roughness (Ra)

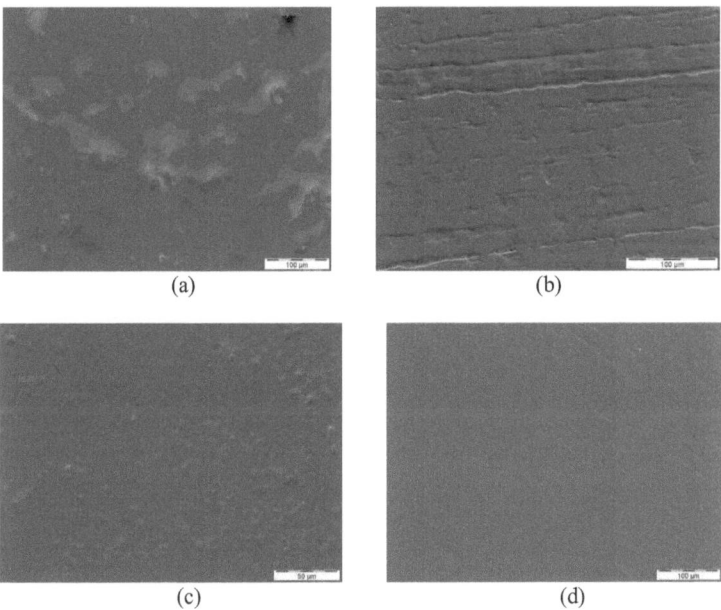

Figure 4.2.1.12: SEM images of samples spin-coated at 1400 rpm using PEI/DMAc (15/85) on substrates treated with (a) acetic acid, (b) 14 HF, (c) HNO₃ and (d) ground.

These morphologies have direct influence on the protective properties of the coatings, as can be seen in figure 4.2.1.13, where the low frequency impedance of HF-treated, acetic and nitric acid cleaned samples are shown. Comparing the values in figure 4.2.1.13, of coatings prepared using PEI/DMAc (15/85), with figure 4.2.1.6b, it can be observed that the acid treated coated samples have a worse performance than the ground ones. In the case of HF-treated substrates, after 96h of exposure time they showed impedance close to that of the uncoated one, in the order of 10^5 Ω cm^2, while the ground substrates showed such impedance only after 240 h of exposure. For the HF-treated substrates coated using PEI/NMP (15/85), the impedance was close to that of the uncoated substrate after 20 h of exposure, and after 48 h, the impedance was even lower, showing that the substrate started to degrade.

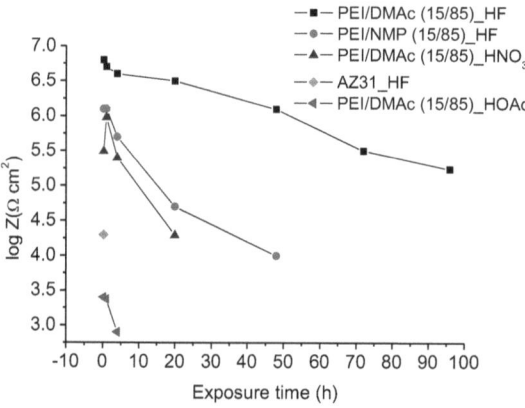

Figure 4.2.1.13: Low frequency (20 mHz) impedance of coated substrates pre-treated with HF and acetic acid (AA) at different exposure time to 3.5 wt.-% NaCl solution.

A very similar behaviour is observed for this sample and the one pre-treated with nitric acid and coated with a DMAc solution. In this case, the main factor associated to the coating performance is not the surface roughness but rather the chemical composition of the interface as will be discussed later. The coatings prepared on acetic acid cleaned substrate showed the worse corrosion behaviour. Despite the impurities removal [1.33, 1.34] this treatment significantly increases the surface roughness which induces defect formation in the coatings.

4.2.2 – Dip-coated poly(ether imide)

4.2.2.1 – Coating characterization

It can be observed in figure 4.2.2.1a that the coatings prepared by the dip-coating method have a dense morphology. Despite the presence of air humidity, the high temperature of the drying step induced a phase inversion process governed by solvent evaporation, resulting in a dense non-porous coating morphology. However, close to the substrate edges the coatings showed some cracks as presented in figure 4.2.2.1b. These defects are formed due to the solution meniscus on the substrate during the coating process that induces lower thickness at the substrate borders. The thickness of the prepared coatings is given in table 4.2.2.1. As expected, there was a thickness variation along the vertical axis of the sheets. As soon as the sheets were removed from the solution the gravity forces the solution downwards, inducing a higher thickness on the inferior part of the substrate. The coating thickness at the superior and inferior parts increased with solution concentration due to viscosity increase (tables 4.2.1.1 and 4.2.2.1). A higher viscosity leads to a higher viscous drag which acts in the opposite direction of the gravity force and results in lower solution outflow and higher thickness. The coatings prepared using 20 wt.-% solutions presented higher thickness variation compared to samples coated with 15 wt.-% suggesting that the higher the solution concentration (viscosity) the higher the coating thickness variation for a specific solvent polymer system. These results demonstrate that the coating thickness can be effectively controlled by solution concentration and therefore by viscosity

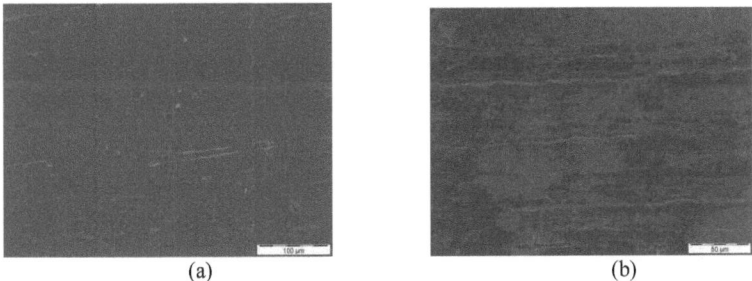

Figure 4.2.2.1: (a) Surface of a sample coated with PEI/DMAc (15/85); (b) higher magnification of the border of the same coating.

Table 4.2.2.1 Thickness variation of the coatings prepared by dip-coating.

Solution	Thickness at superior part (µm)	Thickness at inferior part (µm)
PEI/DMAc (10/90)	approx. 2	4.0 ± 0.5
PEI/NMP (10/90)	approx. 2	4.1 ± 0.8
PEI/DMAc (15/85)	4.2 ± 0.9	4.3 ± 0.3
PEI/NMP (15/85)	3.0 ± 0.3	4.4 ± 0.3
PEI/DMAc (20/80)	11.4 ± 1.2	18.1 ± 2.0
PEI/NMP (20/80)	10.4 ± 1.1	15.7 ± 2.2

Table 4.2.2.2 shows that the residual solvent amount was slightly higher in the coatings prepared using NMP than in the ones prepared using DMAc for both analyzed concentrations. As the difference was lower than 1%, thicker (approx. 160 µm) and free-standing films were prepared in the same manner by both solutions and the residual solvent amount was measured. This was done to confirm that, when films are prepared in the same conditions by NMP and DMAc solution, the one prepared by NMP will have higher residual solvent amount. Table 4.2.2.2 shows that the residual solvent amount off the film prepared by NMP was 1.2% higher compared to the one prepared by DMAc. This is related to the lower vapour pressure of NMP and to its stronger interaction with the polymer [4.19].

To investigate how the presence of residual solvent influences the polymer physical-chemical properties, the glass transition temperature (T_g) of the coatings was evaluated by differential scanning calorimetry (DSC). The T_g is related to the degree of freedom of the polymer chains. Residual solvent can act as plasticizers, which means that it can decrease the T_g and promote the diffusion of water and ions through the coating. In a typical DSC analysis a first heating is performed until a certain temperature to eliminate all volatile components which can interfere in the detection of T_g. This first heating is also used to eliminate the thermal history of the polymer, which stands for the fabrication process of the material. After that the sample is cooled down until a certain temperature below the T_g (in case of semi-crystalline polymer care must be taken to perform the cooling run in the same rate (K min^{-1}) of the first heating to avoid changes in the crystallinity degree). A third run is then performed for the detection of the glass transition temperature and the intrinsic polymer properties.

In the present case, the residual solvent should not be completely eliminated since that its influence on the T_g is the subject of investigation. However, the solvents are eliminated at the same temperature range of the polymer T_g and difficult to analyze. To overcome this problem the following strategy was adopted: a first heating run (10 K min^{-1}) was performed from room temperature to 250 °C to eliminate excess of solvent. This excess of solvent refers to solvent molecules weakly bonded to the polymer. After this process, certain amount of solvent still remains on the film due to the strong interaction with the polymer [4.19]. Then, the cooling run was performed at the same rate (10 K min^{-1}) and the T_g was obtained from this cooling rate. The T_g was not obtained from a second heating run to avoid the complete elimination of solvent. The same process was performed with pure PEI for comparison and the results are shown in table 4.2.2.2

It can be observed that the T_g in both films was lower than that of pure PEI confirming the presence of residual solvent in the film after the first heating. It can also be observed that the T_g of the film prepared using NMP solution was lower compared to the one prepared using DMAc solution. This result is in agreement with the higher residual amount of the film prepared by NMP (which would produce a higher plasticizing effect) and with the worse performance of the coatings prepared with this solution, since that the lower T_g enhances water and ions diffusion. It is important to mention that residual solvent was observed in the coatings even after drying for 24 h in a vacuum oven (10 mbar) at 130 °C indicating strong interactions between these polymer/solvent systems. The chemical nature of such interaction was investigated by FT-IR spectroscopy. Figure 4.2.2.2 shows the infrared spectra of the coatings and of a pure PEI film, The significant influence of the solvents can be observed. In the spectrum of pure PEI the asymmetrical stretching of the carbonyl groups (1725 cm^{-1}) is more intense than the symmetrical one (1779 cm^{-1}) while they have similar intensities for both PEI-solvent systems, besides the presence of at least two new carbonyl signals.

Table 4.2.2.2: Thermo analyses results.

Sample	Residual solvent content (%)	T_g (°C)
PEI/DMAc (15/85)	5.95	-
PEI/NMP (15/85)	6.17	-
PEI/DMAc (20/85)	6.53	-
PEI/NMP (20/85)	6.96	-
Film-DMAc	14.11	187
Film-NMP	15.34	175
Pure-PEI	-	209

According to Cheng et al. [4.22] the signal of the carbonyl asymmetrical stretching only has higher intensity than the symmetrical one when both imide rings are in the same plane, due to an addition of dipole moments. When both imide rings are not in the same plane the intensity of the symmetrical and asymmetrical stretching becomes similar. Therefore, it can be conclude that the presence of both solvents in PEI coatings induces conformation changes in the molecule. The changes in signals between 1200 cm^{-1} and 1300 cm^{-1} suggested a distortion on the ether linkage angle (Ar-O-Ar), as reported by Kostina et al [4.23]. Based in these results, a possible interaction between the polymer and solvents is shown in figure 4.2.2.2a. The new carbonyl signals in the spectra of PEI-solvent system are related to the solvents and the carbonyl interactions shown in Figure 4.2.2.2a.

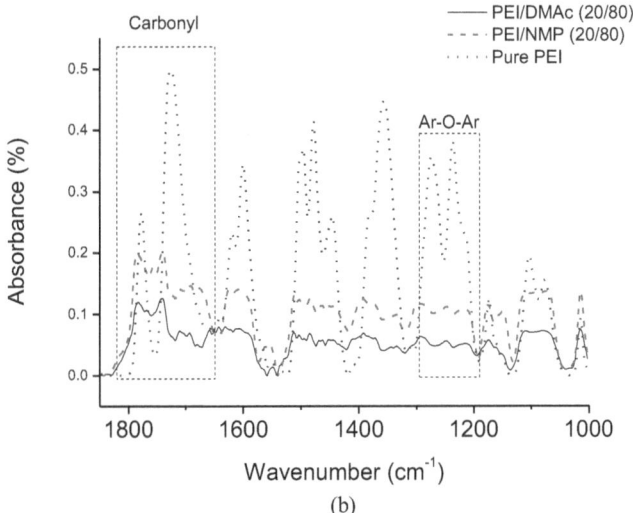

Figure 4.2.2.2: (a) possible interactions between the polymer and the solvents. (b) Infrared spectra of PEI coatings prepared using PEI/DMAc (20/80) and PEI/NMP (20/80) and of one PEI film with low residual solvent content.

4.2.2.2 – Electrochemical impedance spectroscopy.

Figure 4.2.2.3 shows the EIS spectra of coatings prepared using 20 wt.-% solutions, the concentration which resulted in the best performance. It can be observed that the solvents did not have a considerable effect in the initial impedance, and both coatings showed impedance in the order of 10^9 Ω cm^2 in the first hours of exposure to the corrosive solution. Using 10 wt.-% and 15 wt.-% solutions, the coatings showed good initial impedance (in the order of 10^8 Ω cm^2 and 10^9 Ω cm^2, respectively) but after 150 h of exposure these values dropped to $10^6 \, \Omega \, cm^2$ and $10^7 \, \Omega \, cm^2$, respectively. The inferior long term stability for these coatings is related to their lower thickness, as shown in table 4.2.2.1. Despite the similar initial impedance of the coatings prepared by DMAc and NMP as solvents, at longer exposure times they showed distinct behaviours. The coating prepared using DMAc show very good stability, maintaining the initial impedance even after 1000 h of exposure to the corrosive solution, while the coating prepared by using NMP showed gradual impedance decrease after 144 h of exposure reaching $10^7 \, \Omega \, cm^2$ at 336 h. The PEI/DMAc coating showed such impedance only after 1992 h. The thickness of the measured area was the same for both coatings, approx. 13 µm.

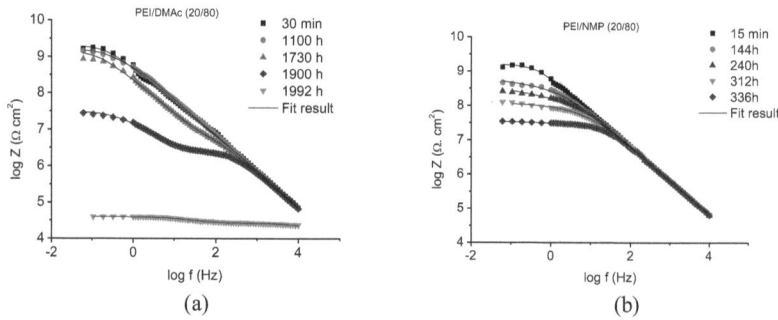

Figure 4.2.2.3: Long term analyses of ground substrates dip-coated using: (a) PEI/DMAc (20/80), (b) PEI/NMP (20/80) at different exposure time to a 3.5 wt.-% NaCl solution.

This difference is related to residual solvent contents in the coatings and to its influence on the polymer T_g, as previously described. Figure 4.2.2.4 shows that the capacitance of the NMP coatings increases faster than that of DMAc one, indicating a faster water uptake rate. This coating capacitance reaches a maximum at 0.34 nF.cm^{-2} and then decreases to 0.24 nF.cm^{-2} at longer times. This decrease is probably related to solvent been washed out from the coating. It is interesting to observe that the DMAc coatings also show a capacitance decrease, but in the first hours of exposure. This is in agreement with the hypothesis of solvent been washed out, since that NMP has stronger interaction with PEI than DMAc [4.19] and would be removed at longer exposure times. Further, DMAc has higher miscibility with water than NMP, which induces a faster removal [4.24]. Besides that, both solvents have considerably high dielectric constants (38 for DMAc and 32 for NMP [4.8]) with significant contribution for the film capacitance. This would produce a detectable capacitance decrease in case of removal.

Therefore, besides the initial lower residual solvent amount in the DMAc coatings, a part of the DMAc is washed out in the first hours of exposure. This process decreases the plasticization of the film and the diffusion of water towards the substrate. For the NMP coatings the solvent is only removed when the electrolytes had already reached the interface and corroded the substrate. Based on these results, it can be concluded that the best performance of the coatings prepared using DMAc is related to the lower amount of residual solvent, that results in a lower plasticizing effect of the polymer and lower diffusion rate of water. It is important to remark that, despite the presence of residual solvent, the EIS behaviour of theses samples could be simulated using the traditional circuits. A detailed discussion about this will be given in the chapter 5.

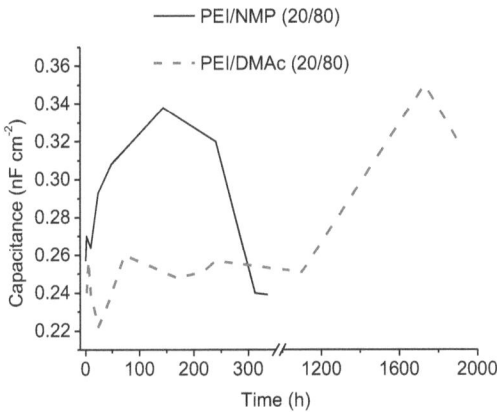

Figure 4.2.2.4: Coatings capacitance changes with exposure time.

4.2.2.3 – Influence of substrate pre-treatment

In section 4.2.1.3, it was shown that the corrosion protection of PEI coatings prepared by the spin-coating method was extremely dependent on the substrate pre-treatment due to roughness effect, which interfered in the formation of a defect free coating. The effect of substrate surface roughness on the morphology of thicker coatings will be discussed in the following. Figure 4.2.2.5 shows the impedance at low frequencies of as-received and acetic acid treated substrates coated with PEI using DMAc solutions in different concentrations, which result in different coating thicknesses. It can be observed that the initial impedance as well as the long term stability of the coating increases with the coating thickness. For as-received substrates with 5 µm of coating thickness the initial impedance was only in the order of 10^4 Ω cm² and after 24 h of exposure the value decreased to the same range of the uncoated sheet. With a coating thickness of 8 µm the initial impedance was in the order of 10^8 Ω cm², just one order of magnitude lower than that of ground substrates with same thickness, but still showed unsatisfying long term stability reaching impedances in the order of 10^5 Ω cm² after only 48 h of exposure to the corrosive solution. At a thickness of 13 µm the initial impedance was also in the order of 10^8 Ω cm² but the long term stability was better, with slightly impedance decrease 72 h of immersion.

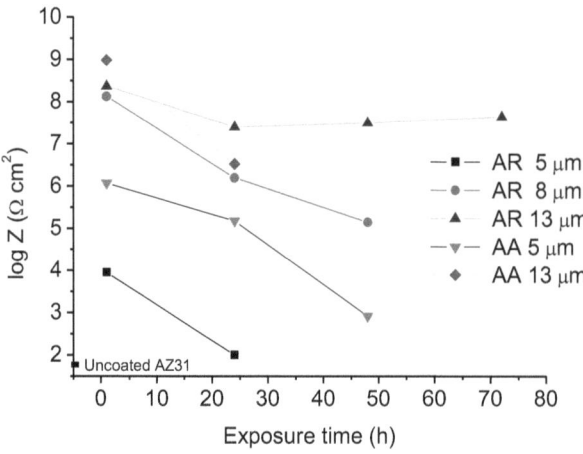

Figure 4.2.2.5: Variation of the impedance at low frequency (20 mHz) as a function of exposure time, for coated samples with different coating thickness, prepared with DMAc solutions. In the figure, AR means as-received and AA acetic acid treated substrates.

The low impedance of as-received coated substrates, especially with coating thicknesses lower than 13 µm, is related to irregularities in the coatings formed by substrate roughness, and to impurities on the substrate surface. Figure 4.2.2.6a show a SEM image of an as-received coated substrate (thickness 8 µm) where roughness induced irregularities can be observed, which decrease the protectiveness of the coating. However, by increasing coating thickness, these irregularities are covered and the impedance values increase too. Nevertheless, the presence of impurities on the surface of as-received substrates results in insufficient long term stability and after 72 h of exposure the coatings were considerably degraded.

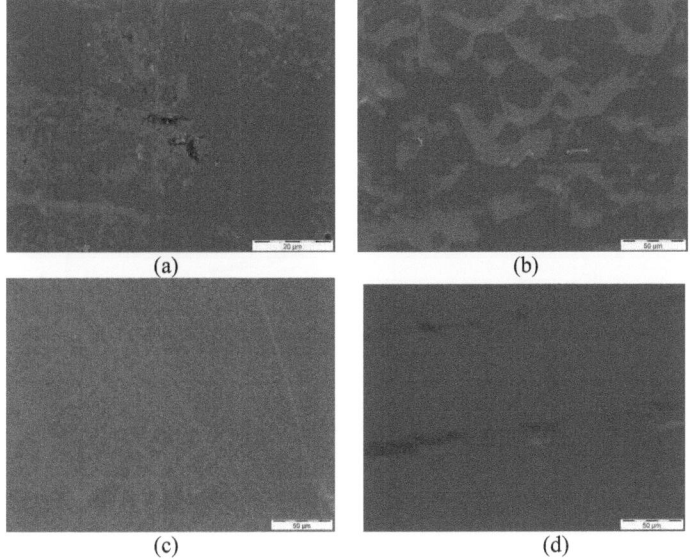

Figure 4.2.2.6: SEM images of the surface of coatings with thickness of *ca.* 8μm prepared on different substrates: (a) as-received AZ31, (b) acetic acid treated AZ31, (c) HNO_3 treated AZ31 and (d) HF-treated AZ31.

For acetic acid treated substrates, which have a roughness of 2.21 μm, even with a coating thickness of 13 μm the corrosion protection considerably decreased after 24 h of exposure to the corrosive solution, as shown in figure 4.2.2.5. The coating surface was very irregular as shown in figure 4.2.2.6b depending on the substrate surface roughness. This result shows that the acetic acid cleaning is not an appropriated pre-treatment for magnesium AZ31 alloy for a coating thickness of about 13 μm or less. The performance of the acetic acid treated samples was even worse than that of the as-received substrates with same coating thickness, showing the relevance of the substrate surface roughness. Acids that remove impurities and do not increase too much the surface roughness are the most appropriate ones to be used for a pre-treatment.

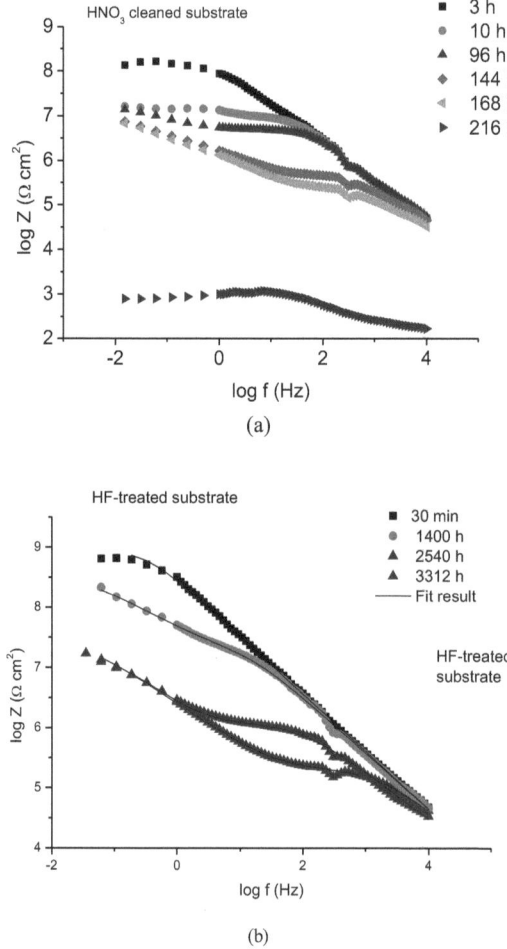

Figure 4.2.2.7: (a) Impedance spectra of a HNO_3-treated substrate dip-coated using PEI/DMAc (20/80) at different time of exposure to the corrosive solution. (b) Impedance spectra of a HF-treated substrate dip-coated using PEI/DMAc (20/80) at different time of exposure to the corrosive solution.

Following these conclusions, figure 4.2.2.7 shows the impedance spectra of coated samples pre-treated with the most appropriate acids: HF and HNO_3. Figure 4.2.2.7a shows the performance of coated HNO_3 cleaned substrate, which shows impedances in the order of 10^8 Ω cm^2 even after 168 h of exposure to the corrosive solution. As expected, the increased performance is associated to a smoother coating surface, as observed in figure 4.2.2.6c as well as to the purity of the substrate. Comparing with the performance of the ground substrate the performance of this sample has decreased. Besides the effect of coating irregularities, the osmotic pressure plays a significant role. Osmosis becomes an important process in the performance of the coatings when soluble salts are present at the interface [1.95]. As shown in figure 3.1.2.2 the cleaning with HNO_3 induces the formation of nitrates at the substrate surface, which are soluble in water [1.34]. The presence of nitrates at the interface of the coated substrate produces an osmotic pressure towards the substrate, which results in swelling of the polymer, and consequently, decreases the barrier property of the coating.

The performance of the coated HF-treated substrate is much superior to all other substrates pre-treatments (figure 4.2.2.7b). It can be observed that this samples have a similar behaviour to those of ground substrates (figure 4.2.2.3a), maintaining the initial impedance in the order of 10^9 Ωcm^2 even after 1000 h of exposure to the corrosive solution. However, these samples show outstanding long term stability, with impedances in the order of 10^7 $\Omega\ cm^2$ even after 3312 h of exposure. This high long term stability is associated to an acid-base interaction between the polymer and the metal surface, as could be observed by XPS spectra (figure 4.2.2.8). As reported in chapter 3.1, the HF-treated substrate consists of a mixture of MgF_2, $Mg(OH)_2$, MgO and possibly $Mg(OH)_{2-x}F_x$ on the surface. As these compounds have acid and basic sites, it is expected that they will interact with polar groups in the polymer (e.g. the imide ring).

Figure 4.2.2.8 shows that the binding energy (BE) of the electrons of the Mg 2p orbital is considerably shifted to higher values on the PEI coated HF-treated substrate compared to the uncoated one. This shift was accompanied by a positive shift in the BE of 1s electrons of fluoride (from 684 eV to 687 eV) and oxygen (from 530 eV to 533 eV). This positive shift indicates that the substrate is acting as a base and the polymer as an acid. This observation is in accordance with studies in the literature which show that MgO and $Mg(OH)_{2-x}F_x$ have basic character, and can even be used as basic catalysts in organic reactions [4.13]. One could expect that the substrate would act as an acid and the polymer as a base, due to the possible interaction between the oxygen of the imide group and the magnesium atom. However, the electronic density of the fluoride and oxygen atoms at the substrate probably inhibits this interaction and the basic character prevails.

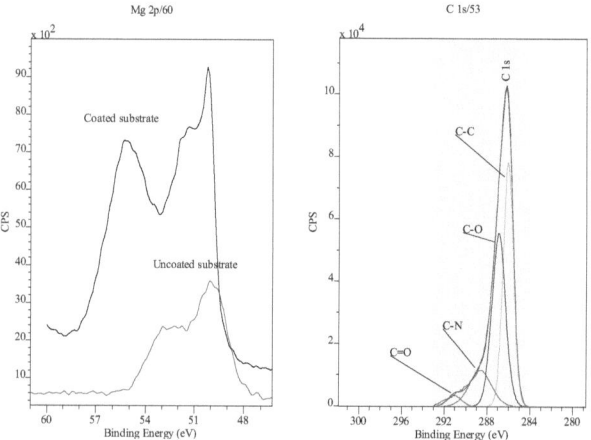

Figure 4.2.2.8: (a) XPS spectra of Mg 2p electrons of the interface of a HF-treated substrate and the substrate coated with PEI-NMP (b) XPS spectra of C 1s electrons of the interface of a coated HF-treated substrate.

The acid site in the polymer structure is not clear due to the complexity in attributing the signals shown in figure 4.2.2.8b to all carbons (including from the residual solvent) shown in figure 4.2.2.2a. To clarify the interfacial interaction in absence of residual solvent, coatings were prepared using CH_2Cl_2 solutions, a solvent that would be almost completely removed by drying at 115 °C. In this case, a positive shift of 1.03 eV was observed for the carbonyl carbon as well as a positive shift in the nitrogen. This suggests the formation of the positively charged nitrogen (see figure 4.2.2.2), since the carbonyl group would be positively shifted in relation to the uncharged structure. In this structure, the positively charged nitrogen is possibly the acid site. In case of NMP based coatings a negative shift of 1.04 eV was observed in the carbonyl signal (which can be related either to the imide ring or to the solvent) and no significant shift for the nitrogen BE could be observed. This suggests that the carbonyl carbon of the uncharged structure is the acid site. In case of DMAc based coatings, small positive shifts in the carbonyl carbon (around 0.2 eV) were observed. Thus it can be concluded that the acid site can be either the positively charged nitrogen or the carbonyl carbons depending on the influence of the residual solvent.

Figure 4.2.2.9 shows that the capacitance of the PEI-DMAc coating on HF-treated substrate is much higher than on ground ones (figure 4.2.2.4). This indicates higher amount of residual solvent, suggesting that the HF-treated surface interacts with the solvent. Part of the solvent is washed out in the first 500 h of exposure, however, the capacitance maintains a high value (0.58 nF.cm^2) compared to the ground substrate (0.25 nF cm^2). After this initial decrease, the capacitance increases constantly, reaching 2.2 nF cm^2 after 3312 h of exposure. This results show that the barrier properties of the coatings are better on the ground than in the HF-treated substrates. Thus, the high stability of the HF-treated coated systems is entirely related to the interface interaction between substrates and polymer-solvent systems.

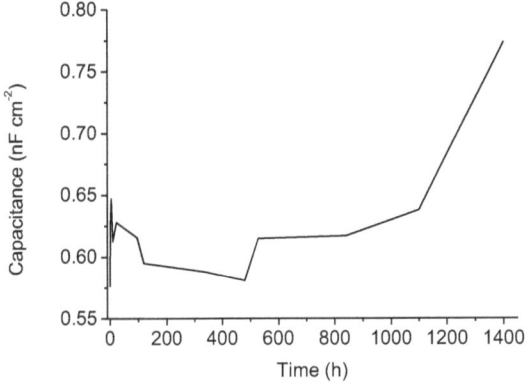

Figure 4.2.2.9: Capacitance change of PERI coating on HF-treated substrate at different exposure time to 3.5 wt.-% NaCl.

It was shown in section 4.2.1.3 that the performance of spin-coatings (2.5 µm of thickness) on HF-treated substrates was inferior to that on ground substrates. This shows that, despite the positive interactions at the interface, the coatings should not be too thin to avoid defect formation by the substrate surface roughness. The performance of the HF-treated substrate starts to be superior to that of the ground substrate when the coating thickness is approx. 8 µm. These results suggested that, for an HF-treated magnesium sheet, a polymeric coating must have a minimum thickness of 8 µm to provide good corrosion protection.

Figure 4.2.2.10 shows the aspect of coated samples after immersion to 3.5 wt.-% NaCl solution. In figures 4.2.2.10a and 4.2.2.10b it can be observed that the corrosion of as-received coated substrates starts in specific points on the surface while for ground coated substrates it starts mainly close to the sample edges and spreads over the sample from that. This different corrosion spots on the surface of as-received coated substrates are related to the impurities and coating defects, as shown in figure 4.2.2.6a. In the case of ground substrates, the corrosion starts mainly close to the substrate borders due to the lower thickness at this part and due to the presence of some defects at these areas, (figure 4.2.2.1b) as discussed on section 4.2.2.1.

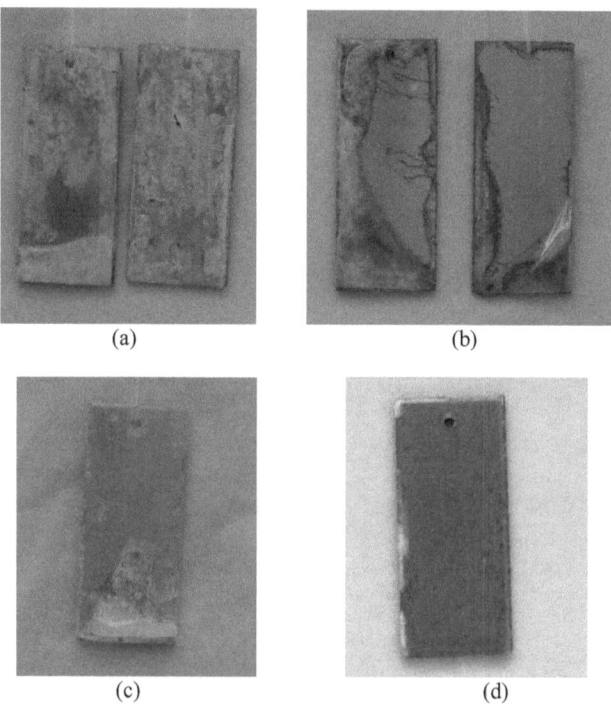

Figure 4.2.2.10: Appearance of the samples coated with PEI/DMAc (20/80) after immersion in 3.5 wt.-% NaCl: (a) as-received substrates after 48h of immersion (b) ground substrates after 48h of immersion (c) HNO_3 treated substrate after 5 days of immersion (d) HF-treated substrates after 7 days of immersion.

After water and ions reaches the polymer/metal interface by this coatings defects at the borders, the coating adhesion decreases and the electrolytes are able to spread along the polymer/metal interface. In some cases, considerable delamination occurs and the corrosion spread easily on the entire sample surface.

It is important to remark that the behaviour in immersion tests of ground substrates coated with PEI/DMAc (20/80) was inferior to those coated with PEI/DMAc (15/85). This shows that, despite the thickness increase, the formation of edges defects could not be avoided, and in fact, could increase due to some stress formation during drying of highly concentrated polymer solutions. This result is very important as it shows that the electrochemical performance of PEI coatings extends by increasing thickness, but the performance in immersion tests can decrease with thickness. Therefore, an appropriated coating thickness must be selected depending on the final application of the coated material.

For HF-treated substrates similar results in immersion tests were obtained with coatings prepared using 15% and 20 wt.-% solutions. The coated sample showed excellent long term stability in immersion tests and even after 7 days of immersion only a small quantity of corrosion products could be observed on the edges, (figure 4.2.2.10d). This result demonstrates that even in the presence of defects on the coating edges, the interfacial interaction is strong enough to protect the residual metal.

Aiming to get more information on this interfacial stability, the dry and wet adhesion of the coatings prepared over HF-treated and ground substrates were determined and the results are given in table 4.2.2.3. It can be observed that the dry adhesion is similar for both polymer-solvent systems and substrates with an average value of 5.5 MPa. However, this value considerably decreases after 12 h of exposure to distillate water, showing the high influence of water in the interfacial stability of these coatings. Similar adhesion values were obtained for all the other substrates. For the DMAc coatings, there was no difference in the wet adhesion for the coatings on ground and HF-treated substrates, while the NMP coatings on HF-treated substrate showed higher adhesion than for the ground ones.

This similar wet adhesion for DMAc coatings on both substrates is surprising, since the HF-treated coated substrate showed much better stability than the ground one. It is possible that the pull-off adhesion test is not sensitive enough to detect a difference in adhesion for these two systems. In case of NMP coatings, the observation of higher wet adhesion for the HF-treated substrate indicates strong interactions, which probably are related to the stronger interaction in the PEI-NMP system than in the PEI-DMAc one. Nevertheless, despite this higher interfacial stability, PEI/NMP coatings on HF-treated substrate had less performance in EIS behaviour than the PEI/DMAc ones due to their high amount of residual solvent.

Table 4.2.2.3: Results of the adhesion tests.

Coating Process	Dry adhesion (MPa)	Wet Adhesion (MPa)
PEI/DMAc (15/85) ground	5.38 ± 0.91	1.79 ± 0.31
PEI/NMP (15/85) ground	5.41 ± 0.57	1.39 ± 0.42
PEI/DMAc (15/85) HF	4.63 ± 1.50	1.62 ± 0.30
PEI/NMP (15/85) HF	6.12 ± 0.180	2.58 ± 0.28

The different interfacial stability of DMAc and NMP coatings on HF-treated substrates suggests that the solvents participate in the interfacial process. The presence of residual NMP enhances the interfacial interaction with the substrate, possibly by providing another acid site which could be the carbonyl carbon of the solvent, as suggested by the XPS analysis. This possible interaction between NMP and the HF-treated substrate results in high amounts of residual solvent and induces higher water and ion diffusion rates. NMP apparently provides a more stable interface for the PEI/HF-treated systems while DMAc provides lower diffusion coefficient of water.

Another remarkable observation which can be made by comparing the results of impedance and immersion corrosion tests is the different performance of ground and nitric acid cleaned coated samples. The performance of ground substrates was much superior to that of nitric acid cleaned ones in EIS tests, but inferior in immersion tests, as can be seen in figure 4.2.2.10b and c. This result is probably related to the presence of impurities in the edges of ground samples. In case of nitric acid cleaning the entire sample is cleaned, while the grinding cleans only the surface. The presence of impurities in the edges of the ground substrates enhanced the coating delamination and decreased its protectiveness. This result demonstrates clearly the role of impurities in the coating deterioration rate.

4.2.3 – Spin-coated PVDF

The morphology of PVDF coatings prepared by spin-coating under nitrogen atmosphere is shown in figure 4.2.3.1 It can be observed that, even in an environment with low air humidity, the coating surface is completely porous. This result is associated to the high hydrophobicity of this polymer which precipitates even in the presence of low relative humidity. A similar coating morphology was obtained by all the tested solutions and after different times of flushing the chamber with nitrogen. As expected this porous morphology resulted in insufficient corrosion protection properties and for that reason the study of PVDF coatings was focused in the dip-coating method.

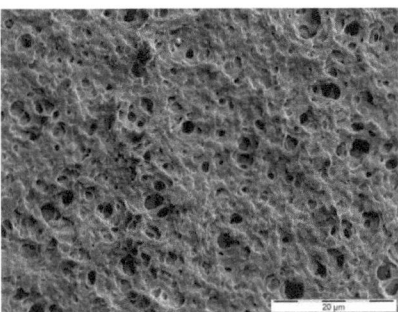

Figure 4.2.3.1 SEM image of a PVDF coating prepared from a PVDF/DMF (15/85) solution by spin-coating under N_2 atmosphere.

4.2.4 – Dip-coated PVDF
4.2.4.1 – Coating characterization

PVDF is a semi-crystalline polymer that can crystallize in different phases, which are α, β and γ [4.25, 4.26]. These different phases have different properties, as the piezoelectric behaviour present in the β phase [4.26] and distinct melting points [4.27]. The melting point and degree of crystallinity of the coatings were measured by DSC analyses using a standard heating, cooling and heating method, all at the same rate of 10 K min^{-1}. The maximum temperature for the heating was 250 °C. The melting temperature was determined by the maximum point of the endothermic peak of the second heating curve. The effect of different solutions on the crystallinity will be discussed in the following chapter and only the specific case of coatings prepared using DMAc solutions will be discussed here as this is representative for all systems.

The crystallinity of this coating was 73% and the melting temperature was 167 °C. This melting temperature is very similar to that of the α phase (170 °C) [4.27] indicating that this is the major crystalline phase in the coating. The presence of other phases could be confirmed by infrared spectroscopy, as shown in figure 4.2.4.1. Besides the typical signals of the α phase, signals related to the β and γ (in small quantities) can also be detected [4.26]. It is important to observe the presence of head-to-head and tail-to-tail defects which were confirmed by the signals at 678, 1330 and 1450 cm^{-1} [4.25]. The intensity of these signals is high in comparison to those in reports of other groups, indicating a considerable amount of these defects [4.28, 4.29]. This is an important character of PVDF films since it has direct influence in the quantity of acidic hydrogen in the polymer back-bone, due to the higher acidity of the hydrogen in the head-to-tail than in the head-to-head configuration.

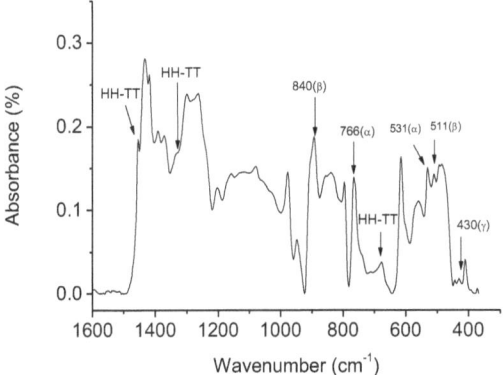

Figure 4.2.4.1: Infrared spectrum of the coating. In the figure, HH-TT means head-to-head and tail-to-tail linkages.

The morphology of the coating prepared using PVDF/DMAc (15/85) solutions is extremely influenced by the drying temperature, as can be observed in figure 4.2.4.2. When the drying temperature is lower than 150 °C there is a considerable influence of the environmental humidity on the polymer morphology and the coating is extremely porous (figure 4.2.4.2a). This is due to a higher rate of polymer precipitation than of solvent evaporation at this temperature. This porous morphology is not only present at the surface but rather at the entire volume of the coating (figure 4.2.4.2b). It forms channels linking the coating surface to the substrate. A similar morphology was observed for drying temperatures of 100, 115 and 135 °C. When the coatings were dried at 150 °C a much denser and homogeneous surface was obtained (figure 4.2.4.2c) with only a few pinholes with diameters ranging from 2 to 5 μm. These pinholes are located at boundaries of the "hills" of the "hills-like" structure of this polymer coating [4.25]. Their distribution at the coating surface was not homogeneous, with considerable areas of the coating without any defect. Figure 4.2.4.2d shows the dense morphology of this coating cross-section. The coating thickness is of approx. 15 μm.

At temperatures higher than the polymer melting point the amount of pinholes in the coating surface increased (figure 4.2.4.2e) suggesting an increase in the crystallinity resulted from the melting-crystallization process. In fact, the size of these pinholes decreases (figure 4.2.4.2e) but its number increases resulting in a high defective coating. Besides that, a decrease in adhesion was observed at drying temperatures higher than 150 °C, which is probably related to a very fast solvent evaporation which prevented good interaction between polymer and substrate. For these reasons the optimal drying temperature was determined as 150 °C, lower than the onset temperature of the melting point (159 °C), to avoid the increase in defects, and high enough to avoid humidity influence.

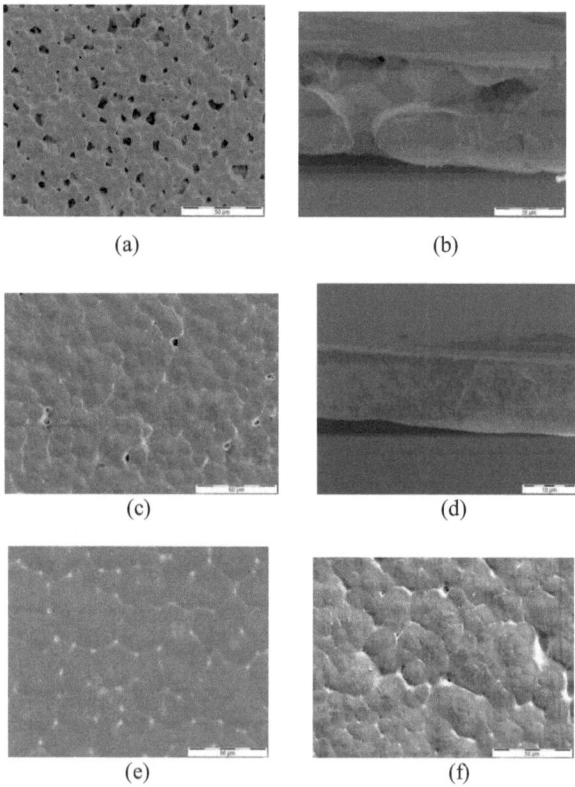

Figure 4.2.4.2: (a) and (b) SEM image of the surface and of the cross-section of a PVDF/DMAc coating dried at 100 °C. (c) and (d) SEM image of the surface and the cross-section of a PVDF/DMAc coating dried at 150 °C. (e) SEM image of the surface of a PVDF/DMAc coating dried at 180 °C (f) SEM image of the surface of a PVDF/NMP coating dried at 150 °C.

Similarly to PEI coatings, the type of solvent did not have considerable influence on the coating morphology, as can be observed by comparing figure 4.2.4.2c and 4.2.4.2f. In case of PVDF coatings even the thickness variation is not affected by the solvent type (table 4.2.4.1). This is an unexpected result due to the difference in viscosity of the solutions with same concentration (table 4.2.4.2). Besides these similarities in morphology and thickness, all of the prepared coatings showed no residual solvent, as observed in TGA analyses (table

4.2.4.1). Due to these and other similarities (see chapter 4.2.4.2) further characterizations were focused on coatings prepared by DMAc solutions. These results are representative for all coatings with other solvents.

Table 4.2.4.1: Thickness variation of the coatings prepared by dip-coating and residual solvent amount.

Solution	Thickness superior part	Thickness inferior part	Residual solvent[*] (%)
PVDF/DMAc (15/85)	5.73 ± 0.46	8.40 ± 0.36	0
PVDF/DMAc (20/80)	11.40 ± 0.73	16.24 ± 0.44	0
PVDF/DMF (15/85)	5.78 ± 0.28	8.03 ± 0.69	0
PVDF/DMF (20/80)	11.57 ± 0.70	15.8 ± 0.56	0
PVDF/NMP (15/85)	6.33 ± 0.45	9.03 ± 0.35	0
PVDF/NMP (20/80)	11.60 ± 0.82	17.00 ± 0.53	0

[*] determined by TGA analyses

For all substrates the adhesion of the coatings was very low except those which were treated with HF. Therefore the measurement of the impedances was difficult. As in the impedance tests the sample is fixed on the cell by a screw (see figure 3.3a), even after total adhesion loss the system still showed good impedance, since the coating was fixed on the substrate by the pressure of the screw. The results obtained in this manner are unrealistic. Due to good adhesion the only realistic results could be obtained from the coated HF-treated substrate, and therefore only these samples will be discussed. The reasons for the good adhesion of this system are explained in details in section 4.2.4.3.

Table 4.2.4.2: Solutions viscosities.

Solution	Concentration (%)	Viscosity (Pa s)
PVDF/DMAc	15	0.406
	20	1.483
PVDF/DMF	15	0.158
	20	-
PVDF/NMP	15	0.951
	20	3.291

4.2.4.2 – Electrochemical impedance spectroscopy

The EIS spectra of the coated HF-treated substrate at different exposure times to 3.5 wt.-% NaCl solution is shown in figure 4.2.4.3a. The EIS behaviour of this sample was fitted using the tradition electronic circuits (figure 1.11) to follow variations in the coating resistance and capacitance. In figures 4.2.4.3a and b it can be observed that, in the first 120 h of exposure the impedance spectra follow a common behaviour, showing a decrease in coating resistance with treatment time. After 6 h of exposure the spectra show two capacitances, indicating the concentration of water at the interface. Despite this expected initial resistance decreases, the coating resistance slightly increases after 120 h and reaches a steady state for the next 2200 h. A similar behaviour is observed for the charge transfer resistance, as shown in figure 4.2.4.3b. The inset in figure 4.2.4.3a clearly shows the impedance increase that took place after 120 h of exposure.

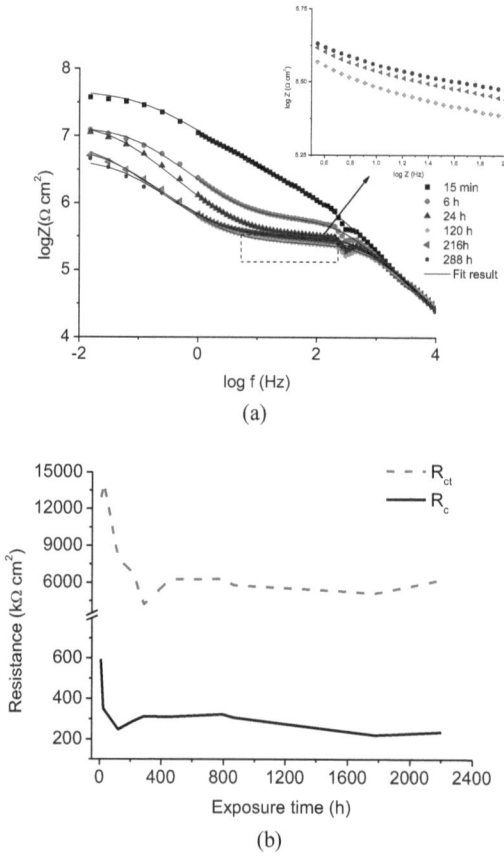

Figure 4.2.4.3: (a) Impedance spectra of PVDF coating on HF-treated substrates. (b) Variation of coating and charge transfer resistance with treatment time.

A similar trend is observed for variations of coating and double layer capacitance, as shown in figure 4.2.4.4. As PVDF is a highly hydrophobic polymer a very low capacitance increase or even a constant value for a certain time was expected. The coating capacitance maintains a nearly constant value after 400 h of exposure but a capacitance decrease is observed in the first hours of immersion. Such capacitance decrease cannot be attributed to solvent removal (as in the case of PEI coatings) due to the lack of residual solvent. It is possible that this capacitance decrease is related to a fill up of pinholes in the coating surface by electrolytes or by corrosion products, which would increase the barrier properties of the film. After the "sealing" of these pinholes, a water back diffusion process could be responsible for the decrease of the coating dielectric constant, and consequently, for the capacitance decrease. Figure 4.2.4.5 shows a SEM image of a coating before and after exposure to the corrosive solution, which demonstrates the deposition of salts or corrosion products in the film surface filling up the pinholes. As soon as all the pinholes in the exposed film are sealed the film capacitance tends to increase very slowly.

Figure 4.2.4.4 shows that the double layer capacitance suffers an intense increase in the first 100 h of exposure and reaches a steady state during the whole exposure time. This initial fast capacitance increase suggests a fast approach and concentration of water at the interface, which probably took place through defects in the coating. Considering this process, the beginning of the steady state represents the point when the majority of the pinholes are filled, maintaining the amount of water at the interface nearly constant. Comparing figure 4.2.4.3 with figure 4.2.4.4 it can be observed that all this parameters reach a steady state at nearly the same time. This suggests the occurrence of certain interfacial processes between polymer and substrate that stabilizes the interface and brings the system to equilibrium. This interfacial process will be discussed in details in chapter 5.3.

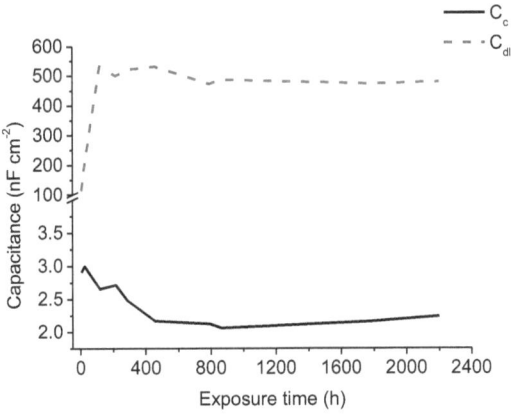

Figure 4.2.4.4: Variations of coating and double layer capacitance with exposure time to the corrosive solution.

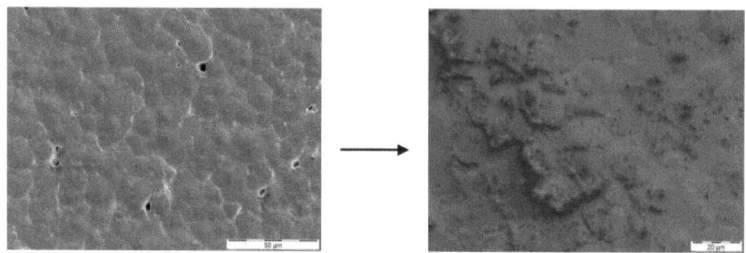

Figure 4.2.4.5: SEM images of a PVDF coating before and after exposure to the corrosive solution.

Comparing figure 4.2.4.4 with figure 4.2.2.4 it can be observed that the capacitance of PVDF coatings is one order of magnitude higher than that of PEI coatings. This might be related to the presence of pinholes and also to the piezoelectric properties of the crystalline β-phase. As described in the literature [4.30], PVDF has an electric dipole moment due to the strong electronegativity of the fluorine atom. When the polymer chains are packed in a crystalline structure in which the dipoles are parallel the cell unit has a net dipole moment. This net moment is strong in the β-phase and inexistent in the α-phase due to its non-parallel arrangement. As shown in section 4.2.4.1 the β-phase is also present in the coating.

Despite this higher capacitance the performance of PVDF coatings was as good as the performance of PEI coatings on HF-treated substrate. The sample maintained high impedances even after 2200 h of exposure to the corrosive solution At this time the experiment was stopped. The sample surface still looked very well indicating that such resistance could have been maintained for a longer time. The performance of the coatings on other substrates and a clear description of the interfacial interaction that renders this superior performance of PVDF coatings on HF-treated substrate will be demonstrated below

4.2.4.3 – Influence of substrate pre-treatment

As previously discussed, the adhesion of the coatings on some substrates was weak and made their analyses by impedance measurements difficult. The effect of substrate pre-treatment on the coating performance could be investigated by immersion tests. In figure 4.2.4.6 the aspect of the coatings in different substrates after immersion tests is shown. It was found that the ground, acetic acid cleaned and as-received substrate showed considerable degradation after 24 h of exposure to the corrosive solution. This is mainly related to the loss of adhesion during immersion. For the ground substrates the dry adhesion is only 0.40 MPa (table 4.2.4.3), a very low value compared to PEI coatings (table 4.2.2.3). With this low adhesion, as soon as water reaches the interface through coating defects the coating is detached from the substrate. Coating delamination was observed for this system after a few minutes of immersion. It is important to mention that the different substrate pre-treatment methods did not produced differences in the coating morphology as they do in case of PEI coatings. The morphology of the PVDF coatings was the same as the ones observed in figure 4.2.4.2 for all substrates.

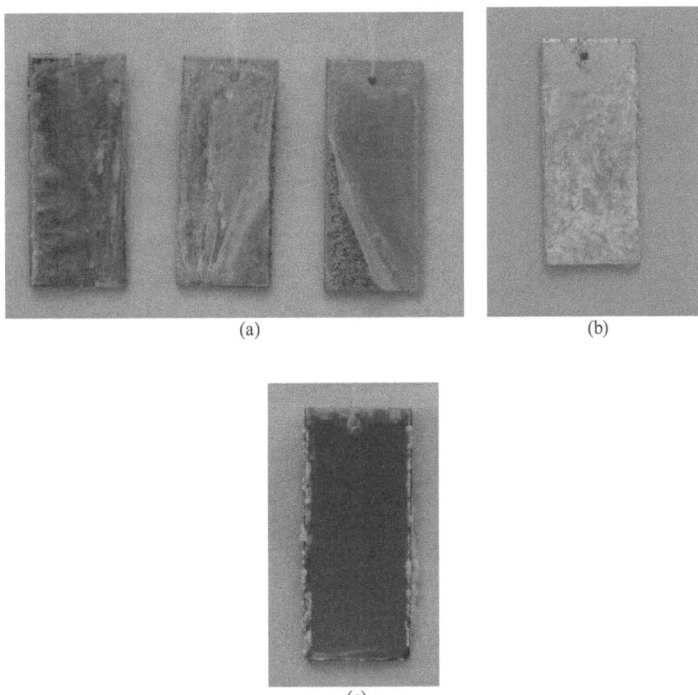

Figure 4.2.4.6: (a) From the left to the right a ground, an acetic acid treated and an as-received substrate coated with PVDF after 24 h of immersion. (b) A HNO_3 treated sample after 5 days of immersion (c) A HF-treated substrate after two weeks of immersion. All tests were performed in a 3.5 wt.-% NaCl solution.

Table 4.2.4.3 Results of adhesion tests.

Substrate	Adhesion (MPa)
Ground	0.40 ± 0.03
As-received	1.37 ± 0.30
HNO_3	0.83 ± 0.35
HF-treated	2.22 ± 1.01
Acetic acid cleaned	0.53 ± 0.11

The samples cleaned by acetic acid showed similar behaviour in immersion and in adhesion tests. The as-received substrate, despite the presence of impurities in the surface showed a slightly better performance than the ground and acetic acid treated ones. This can be related to an interaction between the basic magnesium oxide layer (MgO) and the methylene hydrogen of the polymer backbone. The better performance was accompanied by a better adhesion (1.37 MPa) that also suggests an interfacial interaction. It is important to notice that the substrate roughness has no considerable influence in the adhesion of these substrate/coating systems since that the ground and acetic acid cleaned substrate have different surface roughness but similar adhesion and corrosion behaviour.

Good performance was obtained by the HNO_3 treated substrate in the same way as for PEI coatings (figure 4.2.4.6b). Corrosion products could only be observed at the surface of this sample after 3 days of immersion. In this case, the adhesion strength is not the most relevant parameter as its value is between those of the ground and as-received substrate. A combination of good adhesion (compared to the acetic acid cleaned and ground substrates) and lower impurity concentration at the metal surface plays the most significant role for sample stability.

As expected, the best performance was obtained for the HF-treated substrate (figure 4.2.4.6c). Even after two weeks of immersion this sample showed corrosion products only at the edges and on the superior part, where the coating is slightly thinner. The surface of the substrate was completely free from corrosion attack. The interface of this sample was investigated by XPS spectroscopy and it was observed that the BE of the Mg 2p electrons is shifted to higher values compared to the uncoated substrate (figure 4.2.4.7). This shift indicates that the substrate is acting as a base.

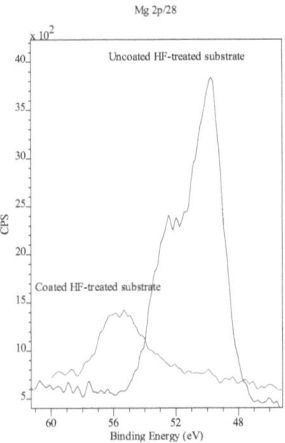

Figure 4.2.4.7: XPS spectra of coated and uncoated HF-treated substrate.

The XPS results show that the acid-base interaction observed for PEI coatings is not a specific characteristic of that system, but rather a common interfacial interaction between the substrate surface and a polymer with acidic groups. Figure 4.2.4.8 presents an model of interaction at the interface between the hydroxides in the metal and the hydrogen of the polymer. This interaction results in a higher adhesion (table 4.2.4.3) but a higher standard deviation is observed for this sample too. This suggests a non-uniform distribution of the basic sites on the substrate surface as described by Wojciechowska [4.31]. It is important to mention that this interfacial interaction will be stronger at a minimum number of head-to-head and tail-to-tail defects because the acidity of the head-to-tail methylene hydrogen is much stronger. Therefore, the adhesion and interfacial stability of this system will be probably higher at a lower number of these defects.

This interfacial interaction is the reason why PVDF coatings perform much better on HF-treated substrate. It renders good adherence for the coating that inhibits the concentration of water at the interface and a high $Mg(OH)_2$ formation rate. The performance of PVDF coatings on HF-treated substrates in immersion tests was even superior to that of PEI coatings. The aspect of PEI coatings after two weeks of immersion was still good but not as excellent as the one shown in figure 4.2.4.6. However, the performance of PEI coatings was inferior to PVDF in other substrates.

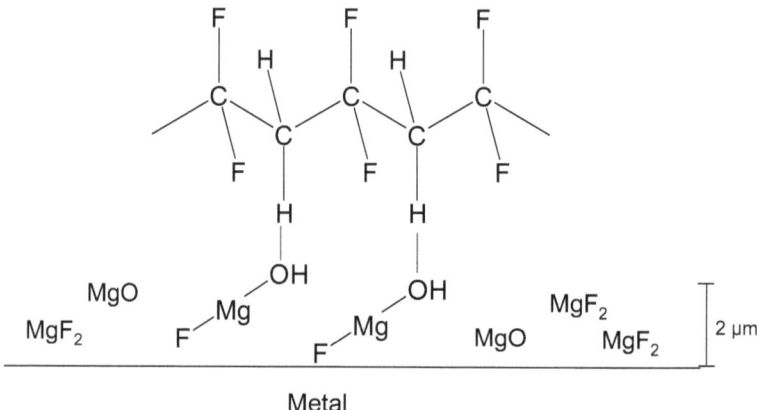

Figure 4.2.4.8: Schematic representation of the interface interaction between PVDF and the HF-treated substrate.

4.2.5 – Spin-coating of polyacrylonitrile

4.2.5.1 – Coating characterization.

Polyacrylonitrile (PAN) is a well known polymer which has been extensively studied in the fields of ultra filtration membranes, textile and carbon fibers [1.88, 1.89]. As a membrane material PAN received considerable interest in the biomedical field due to the easiness in chemical modifying its surface to render good adhesion of biological substances [1.89, 4.32, 4.33]. This easiness in surface modification is related to the presence of the nitrile group in the polymer structure which can easily be hydroxylized in the presence of strong bases. The product of this reaction is usually a salt but acids can also be formed depending on the reaction conditions. Figure 4.2.5.1 shows a general scheme of the chemical modification of PAN on basic environments [1.88, 4.32]. Such modifications enhance the grafting and adhesion of organic molecules as glucose and the immobilization of enzymes, which are very interesting aspects for the preparation of biocompatible implants.

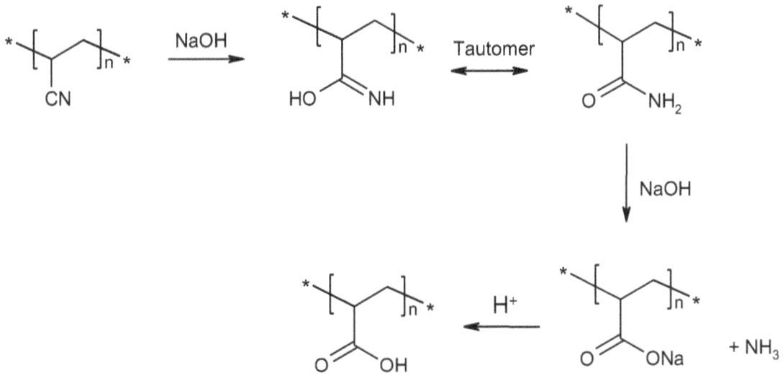

Figure 4.2.5.1: General scheme of the chemical modification of PAN in a basic environment.

For the production of carbon fibers, PAN received attention due to its interesting properties as a precursor [4.34, 4.36]. The process of producing carbon fibers from PAN usually consists in preparing a PAN fiber by the electro-spinning method and then annealing it in an oven at high temperatures [4.36]. During the thermal degradation of PAN the polymer undergoes cyclization and elimination of ammonia and hydrogen cyanide (in absence of oxygen), resulting in fibers with very good mechanical properties. Besides that, PAN fibers prepared by electro-spinning are also applied in textile industries [4.36].

PAN is a semi-crystalline polymer. Its degree of crystallinity is difficult to determine by thermal methods since its degradation temperature is around 300 °C while its theoretical melting point is at 314 °C [4.35, 4.37]. In the present work, thermo gravimetric analyses confirmed the literature values showing degradation temperature of 280 °C, which is related to the cyclising process described above. The only thermal process that could be observed before was the glass transition which takes place at 97 °C. As in this work the study on PAN coatings was aimed to biomedical applications (more specifically, to orthopaedic implants) the spin-coating process is not a suitable method because it does not coat the sample edges and it is restricted to flat substrates. As the entire sample will get in contact with a biological environment a coating method which coats the whole of the sample (as the dip-coating process) is necessary. Nevertheless a brief description on the performance of PAN coatings prepared by the spin-coating process will be provided to serve as a reference to future studies.

The solvent N, N'-dimethyl formamide (DMF) is by far and away the most used solvent for the preparation of PAN membranes and fibers [4.38-4.40]. The properties of PAN membranes and fibers prepared by DMF solution is usually superior compared to the same material prepared by other solvent [4.38]. Such behaviour was also observed in preliminary tests of the present study. Based on these facts, DMF was selected as solvent to prepare PAN coatings. The reasons for the superior properties of PAN when prepared by DMF solutions will be discussed in chapter 5.4. For sake of comparison, the coating was prepared at the same spinning speed (1400 rpm) and with the same thickness as spin-coated PEI (2.5 µm). This thickness could be achieved using an 8 wt.-% solution.

The morphology of PAN coatings prepared by spin-coating was very similar to that of PEI coatings. Figure 4.2.5.2 shows a dense morphology which is obtained when the coating is prepared under dry N_2 atmosphere while a porous morphology is obtained at room atmosphere due to the presence of humidity. In comparison to PEI coatings, the pore size is much smaller for PAN (approx. 0.5 µm). This is due to the lower hydrophobic character of PAN in comparison to PEI. The protective properties of the coating prepared under room atmosphere were poor, and therefore further analyses were performed only for coatings prepared under N_2 atmosphere.

(a)   (b)

Figure 4.2.5.2: SEM image of coating prepared using PAN/DMF (8/92) under (a) N_2 and (b) room atmosphere.

4.2.5.2 – Electrochemical impedance spectroscopy

Figure 4.2.5.3 show the impedance spectra of PAN coatings prepared under nitrogen atmosphere, before and after drying in a vacuum oven at 115 °C. It can be observed that, as expected, the post-dried coating showed a superior performance due to the decrease in residual solvent content. The just spin-coated sample showed very low impedance after less than 48 h while the post-dried coating maintained good impedance during only 4 days. Comparing to figure 4.2.1.6b it can be observed that the performance of PAN coatings is much inferior to that of PEI prepared in the same manner. This is related to a high amount of residual solvent (the just spin-coated PAN had a residual solvent amount of 7.3% and the post-dried of 5.7%). A better understanding of the differences between PAN and PEI spin-coated samples can be obtained by comparing the values of the fitting results shown in tables 4.2.1.2 and 4.2.5.1.

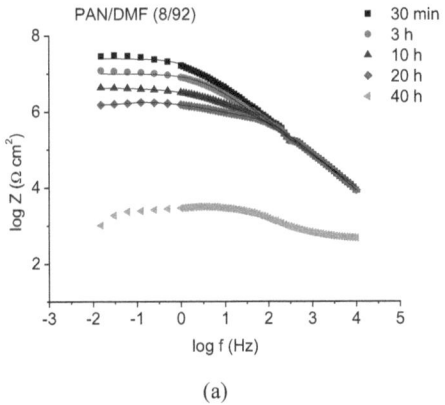

(a)

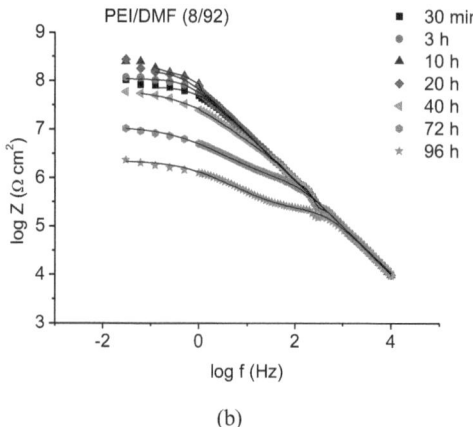

(b)

Figure 4.2.5.3: (a) EIS spectra of an just spin-coated ground substrate at different exposure times to a 3.5 wt.-% NaCl solution. (b) EIS spectra of a post-dried spin-coated ground substrate at different exposure times to a 3.5 wt.-% NaCl solution.

Despite the presence of residual solvent the EIS spectra of PAN coatings could be satisfactory fitted using the tradition circuits, indicating the lack of specific solvent-rich and solvent-poor domains. It can be observed that the values of coatings resistance and capacitance of the just spin-coated and post-dried coatings is very similar to those of PEI (see tables 4.2.1.2 and 4.2.5.1). In general, PAN coatings showed slightly higher capacitances and lower resistances (in comparison to the solvent poor domains of PEI, on the case of just spin-coated samples). This result indicates that the extremely different performances of PAN and PEI are not mainly related to different rates of water diffusion through the coating, which would induce distinct values of T and R. The fact that the post-dried PAN has only slightly superior performance than the just spin-coating corroborates this conclusion. This suggests that the poorer performance of PAN coatings is related to the interfacial stability of the systems.

Table 4.2.5.1: Results of the fitting of the EIS spectra. In the table, T_c and T_{dl} are the T constant of the coating and double layer CPE, respectively.

Exposure time	R_c (Ω cm^2)	T_c (Ω^{-1} cm^2)	R_{ct} (Ω cm^2)	T_{Cdl} (Ω^{-1} cm^2)
Just spin-coated				
30 min	2.5 x 10^7	6.1 x 10^{-9}	-	-
3 h	1.0 x 10^7	9.8 x 10^{-9}	-	-
10 h	7.4 x 10^5	2.9 x 10^{-9}	3.6 x 10^6	5.1 x 10-8
20 h	7.1 x 10^5	2.7 x 10^{-9}	1.0 x 10^6	1.2 x 10^{-7}
40 h	-	-	-	-
Post-dried				
30 min	7.4 x 10^7	2.4 x 10^{-9}	-	-
3 h	1.1 x 10^8	2.5 x 10^{-9}	-	-
10 h	2.0 x 10^8	2.3 x 10^{-9}	-	-
20 h	1.6 x 10^8	2.2 x 10^{-9}	-	-
40 h	3.1 x 10^6	1.7 x 10^{-9}	6.8 x 10^7	8.9 x 10^{-9}
72 h	9.0 x 10^5	1.6 x 10^{-9}	1.0 x 10^7	5.0 x 10^{-8}
96 h	2.1 x 10^5	1.6 x 10^{-9}	2.0 x 10^6	1.5 x 10^{-7}

This assumption was confirmed by adhesion tests. The adhesion of PAN coatings to ground substrates was very similar to that of PVDF coatings, showing values usually below 0.5 MPa. This low adhesion induces a complete detachment of the coating form the substrate as soon as water reaches the interface. As can be observed in Table 4.2.5.1, in case of just spin-coated samples, the water reaches the interface after only 10 h of exposure, which is indicated by the presence of double layer capacitance in the fitting process. The post-dried coating showed better barrier properties, demonstrated by the lower initial value of the coating capacitance, but after 40 h of exposure water could reach the interface. After that, the polymer is easily detached from the substrate and water concentrates at the interface as observed by the fast increase in the double layer capacitance.

The performance of these coatings on other substrates was even worse due to the roughness induced defect formation (see PEI coatings). Thus, it can be concluded that the properties of PAN coatings prepared by spin-coating are not sufficient to protect the metal in the tested environment. The dominant aspect of the insufficient protectiveness of this coating is the lower interfacial stability. It is possible to increase the protectiveness of the coating by increasing its thickness, but as discussed for PEI, other methods are more appropriated for the preparation of thicker coatings. Nevertheless, PAN coatings prepared by the spin-coating method could be an interesting choice to prepare thin and biocompatible top coatings over a protective intermediate layer in multi-layered coatings.

4.2.6 – Dip-coated polyacrylonitrile

4.2.6.1 – Coating characterization.

Table 4.2.6.1 shows the thickness variation and solution viscosities of PAN coatings prepared using two different solutions in DMF. As expected, the thickness increases as the solution concentration passes from 6 wt.-% to 8 wt.-%. It is interesting to observe that the thickness of PAN coatings prepared using 8 wt.-% solutions is superior to the thickness of PEI coatings prepared using 15 wt.-% solutions (see Table 4.2.2.1). This is related to the higher viscosity of the PAN solution (compare tables 4.2.1.1 and 4.2.6.1) which is a consequence of the higher molecular weight of the polymer (see Experimental Part). The residual solvent content in the coatings was around 4.0 wt.-% Similar to PEI, the morphology of PAN coatings prepared by the dip-coating method was dense and non-porous.

Table 4.2.6.1: Thickness variation of PAN coatings prepared by the dip-coating method and the viscosity of the used solutions.

Solution	Thickness at superior part (μm)	Thickness at inferior part (μm)	Viscosity (Pa.s)
PAN/DMF (6/94)	3.5 ± 0.3	5.4 ± 0.1	0.43
PAN/DMF (8/92)	7.5 ± 0.6	10.7 ± 0.4	1.17

4.2.6.2 – Electrochemical impedance spectroscopy

Figure 4.2.6.1 shows the impedance spectra of PAN coatings prepared using 6 and 8 wt.-% solutions on ground substrate. The first has low long-term resistance (figure 4.2.6.1a), and after only 3 days of exposure the impedance drops from 10^7 Ω cm^2 to 10^5 Ω cm^2. This is related to its low thickness as discussed for PEI coatings prepared using 10 wt.-% solutions. The coating prepared using 8 wt.-% solution show a much better performance due to the higher thickness (figure 4.2.6.1b). For that reason, further analyses were only performed on coatings prepared using the solution PAN/DMF (8/92).

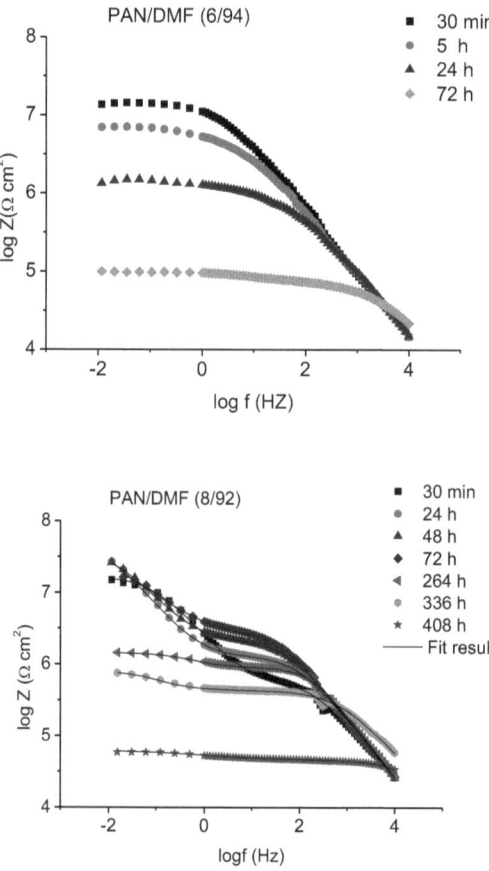

Figure 4.2.6.1: EIS spectra of PAN coatings on ground substrate prepared using the following solutions: (a) PAN/DMF (6/94), (b) PAN/DMF (8/92).

Figure 4.2.6.2 shows the EIS spectra of a HF-treated substrate coated with PAN/DMF (8/92) after different exposure times to a 3.5 wt.-% NaCl solution. For both substrates (ground and HF-treated) there is an increase in impedance with exposure time (in the first 72 h for the ground shown in figure 4.2.6.1b and in the whole measured exposure time for the HF-treated one). This can be related to solvent extraction and to reactions between coating and corrosion products, as mentioned for PEI and PVDF coatings. When water reaches the interface magnesium hydroxide will be formed and it is expected that an interfacial reaction will take place (as the one shown in figure 4.2.5.1). To investigate this, infrared spectroscopy was used. The spectra of the coating on HF-treated substrate, before and after exposure to the corrosive solution, are shown in figure 4.2.6.3. By analyzing the spectra, the reaction between magnesium hydroxide and the polymer is confirmed by the appearance of the signals from 3380 to 3700 cm^{-1} (O-H), at 3250 cm^{-1} (N-H) and by the decrease in the CN/CH signals ratio from 1.67 to 1.24. All this supports the formation of the tautomer structure shown in Figure 4.2.5.1. It is also possible that the formation of the salt and the acid took place, but this is difficult to check due to an overlapping in the signals of the carbonyl by the DMF signal (this signal is observed in many studies on PAN films and is usually regarded as a complex formed between solvent and polymer) [4.40].

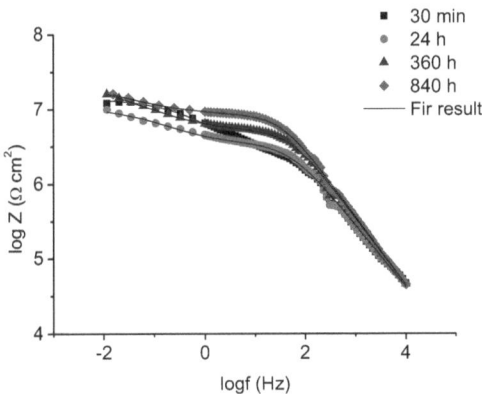

Figure 4.2.6.2: EIS spectra of PAN coating prepared using PAN/DMF (8/92) on HF-treated substrate at different exposure times to a 3.5 wt.-% NaCl solution.

As this reaction leads to new acid sites in the polymer it is expected that the interfacial interaction will increase as the reaction goes on. This is certainly related to the impedance increase observed in the EIS spectra. Besides that, the ratio of DMF/C-H signal (the signal of the tertiary carbon present only in the PAN structure) decreases from 2.80 to 1.44 after immersion, indicating a decrease in the residual solvent content. These two parameters are the more significant ones related to the observed electrochemical behaviour of the samples.

In case of the coating on ground substrate, the lack of a stable basic layer at the interface inhibited the increase in the interfacial stability through acid-base interaction. Due to this reason, after 400 h of exposure the ground coated sample showed impedance in the order of 10^4 Ω cm^2 while the coated HF-treated substrate maintained high impedance even after 800 h of exposure (figures 4.2.6.1b and 4.2.6.2). It is important to mention that if the interfacial reaction proceeds to the formation of the salt, ammonia will be eliminated (figure 4.2.5.1) and will probably induce detachment of the coating due to an increase of the pressure at the interface. Therefore, it is desired that the interfacial reaction stops at the first step.

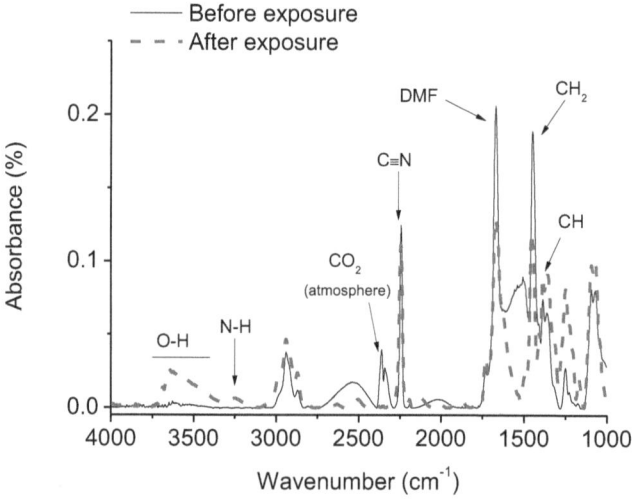

Figure 4.2.6.3: Infrared spectra of the coatings on HF-treated substrates before and after 2 months of immersion in 3.5 wt.-% NaCl.

In figure 4.2.6.4 it is shown the capacitance variation with exposure time of PAN coatings on HF-treated and ground substrate. Different from the ones observed for PEI, the capacitance of the PAN coated HF-treated substrate was inferior to that of the coated ground substrate. This suggests that the interaction of the substrate with DMF is not as strong as its interaction with DMAc. It can be observed from figure 4.2.6.4 that the capacitance of PEI coatings on HF-treated substrate does not increase in the whole measured time, while in ground substrate it shows an increase after 250 h of exposure. The lack of a capacitance increase in case of the coated HF-treated substrate may be related to interfacial interactions/reactions that inhibits the concentration of water at the interface and possibly induces water back diffusion.

Comparing figure 4.2.6.4 with figures 4.2.2.4, 4.2.2.9 and 4.2.4.4 it can be observed that the capacitance of PAN coatings is at the same order as that of PVDF but one order of magnitude higher than that of PEI. Another similarity between PAN and PVDF is the capacitance variation. In both cases, the capacitance decreases in the first 400 h of immersion in a similar magnitude (1 nF cm^{-2} for PVDF and 1.5 nF cm^{-2} for PAN). As previously mentioned, this can be related to solvent been washed out and to the reactions at the interface.

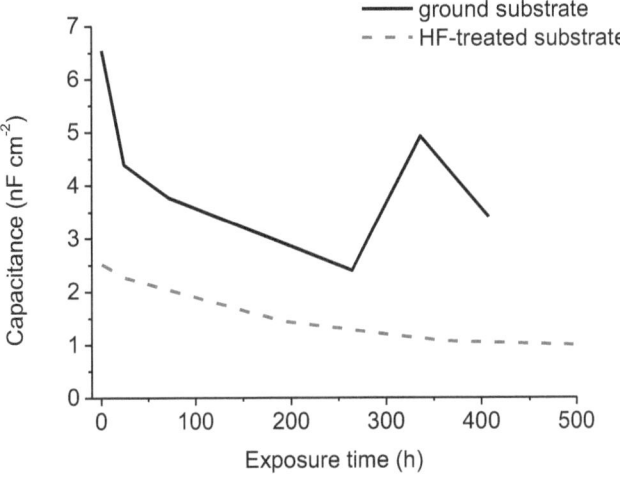

Figure 4.2.6.4: Variation of coating capacitance with exposure time in 3.5 wt.-% NaCl solution.

4.2.6.3 – Influence of substrate pre-treatment

Figure 4.2.6.5 shows the impedance at low frequencies for PAN coatings on as-received, nitric and acetic acid cleaned substrates. It can be observed that the HNO$_3$ cleaned substrate showed the highest long-term stability. The as-received substrate showed the higher initial impedance but poor long-term stability due to well discussed influence of impurities. As expected, the acetic acid cleaned substrates showed the lowest impedance in the initial exposure time, but better long term stability than the as-received one. Both coated substrates (as-received and acetic acid cleaned) were considerably damaged after only 24 h of exposure. This result supports the general conclusion that acetic acid is not a suitable pre-treatment for coating with thickness lower or equal to 10 μm. The good performance of the HNO$_3$ cleaned sample is also in accordance with previous results.

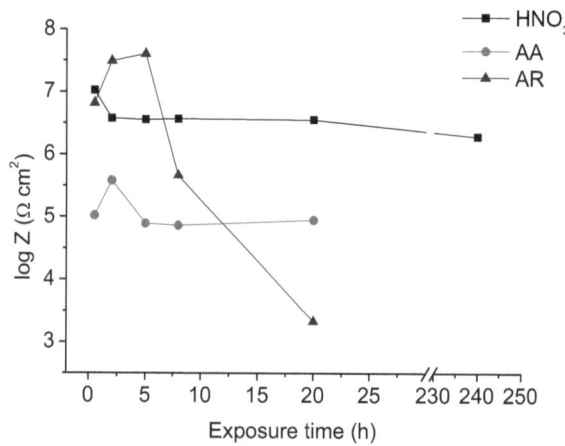

Figure 4.2.6.5: Low frequency (20 mHz) impedance of coated substrates pre-treated with HNO$_3$, acetic acid (AA) and as-received (AR) at different exposure time to 3.5 wt.-% NaCl solution.

Figure 4.2.6.6 shows images of differently pre-treated substrates coated with PAN after immersion in a 3.5 wt.-% NaCl solution. The same trend presented for PVDF is observed. The best performance is obtained with the HF-treated substrates, followed by the nitric acid treated one. The performance of acetic acid cleaned and as-received samples was not satisfying. The reasons for that are the same as those discussed for PEI and PVDF coatings. The performance of ground substrates was also very poor due to the low adhesion of the coating (see Table 4.2.6.2). The adhesion of PAN coatings was very similar to that of PVDF, except for HF-treated and as-received substrate where PVDF coatings have better adhesion. This is probably due to a stronger acid-base interaction between the hydroxides on the metal surface and the hydrogen in PVDF compared to possible acid-base interaction with the nitrile carbon of PAN. The presence of the triple bond and the free electron pair of the nitrogen atom probably inhibit the interaction between the substrate and the nitrile carbon, resulting in weak interfacial interactions and low adhesion properties.

Figure 4.2.6.6: Images of PAN coatings on different substrate after immersion in 3.5 wt.-% NaCl solution: (a) ground substrate 1 day of immersion, (b) acetic acid treated substrate after 1 day of immersion (c) as-received substrate after 3 days of immersion (d) HNO_3 treated substrate after 3 days of immersion and (e) HF-treated substrate after 7 days of immersion.

Table 4.2.6.2: Results of the adhesion tests.

Substrate	Adhesion (MPa)
Ground	0.43 ± 0.01
As-received	0.37 ± 0.12
Acetic acid cleaned	0.74 ± 0.24
HNO_3 cleaned	0.40 ± 0.07
HF-treated	0.75 ± 0.20

The interface of the PAN/HF-treated sample was also investigated using XPS spectroscopy. The spectra are shown in figure 4.2.6.7. The Mg 2p signal has a similar pattern as the ones observed in the case of PVDF and PEI coatings. A binding energy of 56 eV is observed for the interacting magnesium while the non-interacting substrate has a binding energy around 50 eV. As previously commented this indicates that the substrate is behaving as a base. Interesting signals could also be observed at the interfacial C 1s spectra as shown in Figure 4.2.6.7b. It can be observed that a signal at 291 eV appear at the interface which is related to a carbonyl group. The presence of this carbonyl signals indicates that the basicity of the substrate surface is high enough to induce its reaction with the polymer. Similarly to PEI and PVDF, the best performance of PAN coatings on HF-treated substrate compared to other substrate is related to acid base interactions at the interface.

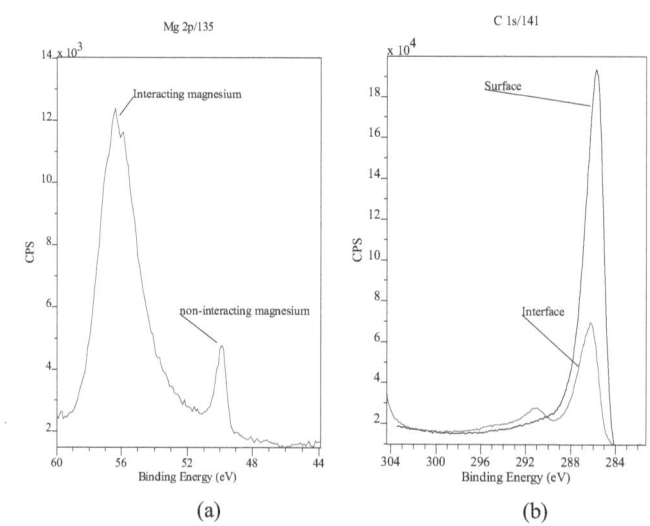

Figure 4.2.6.7: XPS spectra of the interface between PAN and the HF-treated substrate.

Despite these similar interface interactions and patterns of XPS spectra, no similarities were observed in the adhesion of these three polymers on HF-treated substrates. The stability increase produced by this interfacial acid-base interaction is strong enough to render better performance on corrosion tests but not strong enough to render higher adhesion, at least in pull-off tests. Only in some specific cases a higher adhesion could be observed in the HF-treated substrate (PEI coatings prepared using NMP and PVDF coatings, which showed higher adhesion but also higher standard deviation on the measurements). Although the interfacial interactions of PEI, PVDF and PAN with the HF-treated substrate are similar, the substrate was much more damaged after corrosion tests in case of PAN coating. This indicates lower interfacial stability, compared to the other polymers.

4.2.6.4 – Tests on simulated body fluid (SBF)

As previously discussed PAN is a very interesting polymer for biomedical applications due to the high potential to modify its surface for biocompatibility and adhesion of organic molecules. As magnesium implants have received considerable attention as biodegradable implants, as well as HF-treated magnesium alloys due to the positive effect of fluoride in the bone structure [1.42, 1.45], it is a very interesting approach to coat the HF-treated magnesium sheet with PAN and check how the metal corrodes in a simulated body fluid (SBF). The SBF is a solution with a salt composition similar to that of a human body (table 4.2.6.3). The tests were performed at 37 °C using a thermostat bath for better simulation of the biological environment. In figure 4.2.6.8 the impedance spectra of the analyses is shown. By comparing figure 4.2.6.8 with figure 4.2.6.2 it can be observed that, despite the higher initial impedance, the long term stability of the coating on SBF was inferior to that in 3.5 wt.-% NaCl. In SBF the coating impedance dropped to 10^5 Ω cm^2 after 792 h of exposure while it maintain in the order of 10^7 Ω cm^2 after 840 h of exposure to 3.5 wt.-% NaCl solution. As the SBF is milder than the 3.5 % NaCl solution (lower salt concentration) it could be expected that the coating performance would be superior in SBF. This behaviour can be explained considering the osmotic pressure in both solutions

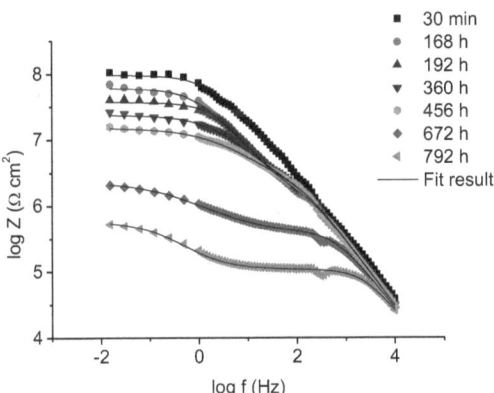

Figure 4.2.6.8: EIS spectra of a coated HF-treated substrate on different exposure times to a SBF at 37 °C.

The osmotic pressure of SBF is 8 mbar while for the 3.5 wt.-% NaCl solution it is 30 mbar (calculated using the following equation P = iCRT, where P is the osmotic pressure; i is the ionic factor; R is the gas constant and T the test temperature). At the beginning of the treatment, the osmotic pressure acts in the direction of the solution (which is more concentrated compared to the interface). In this situation, the diffusion of water trough the coating is much more hindered in the 3.5 wt.-% NaCl than in SBF. This leads to a higher water diffusion rate in SBF. This situation would only change when ions reach and concentrate at the interface, changing the osmotic pressure. However, for all measured samples, no detectable amounts of solution ions were observed at the interface by XPS analyses, even after one month of exposure to the corrosive solution. Thus, during all the exposure time the osmotic pressure on 3.5 wt.-% NaCl is higher, acting in the solution direction, resulting in more hindered water diffusion in this solution, compared to SBF.

In figure 4.2.6.9 it can be observed that in the beginning of the exposure, the coating capacitance in 3.5 wt.-% NaCl solution was higher than in SBF. As the exposure time goes on, both capacitances decrease, but after 456 h the coatings capacitance in SBF experiences a continuum increase. It can observe that the time when the capacitance increase coincides with the time when the impedance considerably decreases (figure 4.2.6.8). The constant decrease in 3.5 wt.-% NaCl shows that the rate of water uptake is lower than the rate of solvent removal in the whole measurement time. On the other hand, in SBF the impedance increase which takes place after 456 h of exposure indicates an increase in the water uptake rate. This result is consistent with the lower osmotic pressure (which acts in direction of the solution) in SBF, which would induce higher water diffusion rate, compared to the 3.5 wt.-% NaCl solution.

Table 4.2.6.3: Composition of the simulated body fluid compositions [4.41].

Salt	Concentration (g L^{-1})
$CaCl_2.2H_2O$	0.185
$MgSO_4.7H_2O$	0.06
KCl	0.4
K_2HPO_4	0.05
$NaHCO_3$	0.35
NaCl	8.0

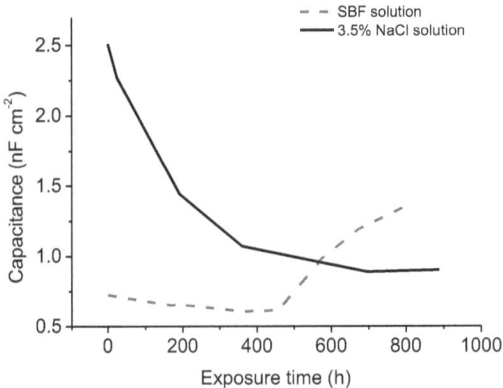

Figure 4.2.6.9: Variation of coating capacitance with exposure time to 3.5 wt.-% NaCl and SBF of PAN coatings on HF-treated substrate.

Besides the electrochemical tests, immersion tests were performed in SBF. Figure 4.2.6.10 shows images of the aspect of the samples at different immersion times. It can be observed that after 9 days of immersion there were considerable amount of corrosion products (mainly hydroxides and carbonates as obtained by XPS analyses) at the interface that resulted in the lost of the usual colour of the HF-treated substrate. A closer look at this sample revealed many little bubbles along the coating, suggesting that water could reach the interface in different spots. A more intense corrosion attack could be observed at the lower edge of the sample. Compared to figure 4.2.6.6e it can be observed that the corrosion process in immersion tests is much more intense in SBF than in 3.5 wt.-% NaCl as observed in the EIS investigation. This is also related to the lower osmotic pressure in SBF solution, which enhances the water diffusion trough the coatings.

After 18 days the sample maintain more or less the same aspect showing that, despite the fast arrival of water at the interface, the sample could maintain stability due to the interfacial process. However, the more server corroded area looked even more damaged and the coating was completely disrupted due to the pressure produced by the increase volume of corrosion products. After 21 days of immersion, the entire lower edge of the sample was corroded and dip pits were formed that could be observed from the other side of the sample. Another sample was investigated in the same way and similar results were obtained.

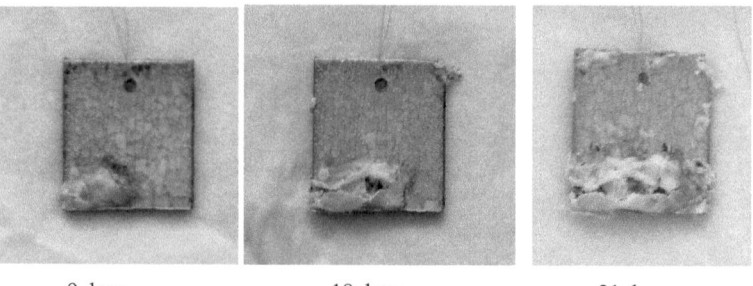

        9 days   18 days   21 days

Figure 4.2.6.10: Image of the aspect of the PAN coated HF-treated substrate after different immersion times in SBF at 37 °C.

      The immersion test is a much more significant test for the evaluation of implants performance than the EIS one, as the whole implant is in contact with the corrosive environment. The required stability for an orthopaedic implant is of at least 2 months, thus it can be concluded that the AZ31 alloy pre-treated with HF and coated with PAN do not have the required stability to be used as an implant. However, the corrosion resistance of the alloy considerably increased by the HF-treatment followed by coating with PAN. Figure 4.2.6.11 show the aspect of an untreated AZ31 alloy sheet after 5 days of immersion in SBF. The sample surface is completely corroded and many pits are observed. After 12 days of immersion, the sample was completely dissolved in the solution. Nevertheless, further tests are required to allow a conclusive description on the potential of PAN as corrosion protective coatings for magnesium AZ31 alloys. Future studies which aim to investigate the potential of this method should focus on biocompatibility of the sample. Tests *in vivo* are important to give insights on the sample behaviour on an *in service* conditions. Besides that, potential risks related to residual amounts of DMF should be considered to allow a conclusive evaluation of the potential application of the adopted strategy.

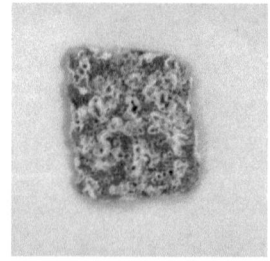

Figure 4.2.6.11: Image of the aspect of an untreated AZ31 alloy sheet after 5 days of immersion in SBF at 37 °C.

5 - Discussion of the results

5.1 – Substrate pre-treatments

The substrate pre-treatment is considered as the most relevant parameter for the performance of corrosion protective coatings, as stated by ISO 8502 which says that "the behaviour of protective coating systems is affected mainly by the condition of the substrate immediately before the coating system is applied". In the past it has been shown that the performance of corrosion protective coatings is more related to an interfacial stability between coating and substrate then to barrier properties [1.95, 1.97]. In other words, even if water and ions are able to diffuse through a coating, they will not be able to produce significant damage to the metal if the interface is stable. In the present work acid cleaning methods were applied due to their good impurity removal effect and the possibility of formation of a protective layer, which is the case of HF acid treatment. The results obtained by the HF treatment as well as by the cleaning with acetic and nitric acid and grinding will be discussed in details.

5.1.1 – HF treatment

The corrosion protection of Mg alloy AZ31 obtained by the HF treatment is mainly related to the formation of a protective layer on the substrate surface. The Fe/Mn ratio is reduced to below its critical value, but this is not the main factor governing the corrosion behaviour since that the aqueous solutions of 14 and 28 mol L^{-1} HF resulted in similar Fe/Mn ratios but different electrochemical performances. The formed layer is mainly constituted by MgF_2 as shown by XPS analyses, but the presence of hydroxides and oxides is also confirmed. The formation of hydroxides (in the form of $Mg(OH)_xF_{2-x}$) in acidic environments is well documented in the literature, despite the fact that its mechanism is not well understood [1.42, 5.1]. Verdier [5.1] suggests that this compound is formed either by simultaneous reaction between the Mg^{2+} and the anions OH^- and F^- or by a substitution reaction, where the hydroxide anions of the magnesium hydroxide film are gradually replaced by fluoride. As in the present study the quantity of hydroxides on the metal surface increases with treatment time (figure 4.1.1.4) a substitution of the fluorides by hydroxides is suggested, instead the one proposed by Verdier. This is consistent with the fact that the signal related to MgF_2 appeared in the initial phase of the treatment, and the signals related to the hydroxides appeared only at the end. The conversion of MgF_2 to $Mg(OH)_2$ is reported in the literature, showing that this reaction is thermodynamically possible [5.2].

Another mechanism is proposed by Wojciechowska [4.31], who observed that when MgF_2 is in contact with water (including water vapour) its coordination shell is filled by hydroxide anions. Wojciechowska shows that, due to the crystalline structure of MgF_2 (rutile-type) some magnesium atoms in specific planes of the crystalline structure are coordenatively unsaturated and act as acid sites, being able to receive a hydroxide anion. Three different hydroxides were identified depending on their location in the crystal structure, and also three different hydroxides were observed in the infrared of the HF-treated samples in the present work (figure 4.1.1.5a). Therefore it is possible that the hydroxides observed in the present work are formed in the way shown by Wojciechowska. After the formation of MgF_2, the contact with water fills the coordination shell of unsaturated magnesium ions with hydroxides. In this way there is no replacement reaction, but rather a filling of vacancies in the lattice of MgF_2 crystalline structure by hydroxides.

The presence of hydroxides renders interesting properties for the substrate surface, as basic character. On the other hand, the magnesium atom in the magnesium fluoride molecule can be considered as acid in the Lewis sense (able to receive an electron pair). The created surface can behave as an acid or as a base in contact with an upper layer, depending on its chemical structure. According to Wojciechowska the basic character of magnesium fluoride is stronger than the acidic one and therefore it would be expected that polymers with acidic groups would strongly interact with the substrate surface. This was observed for all 3 polymers tested in this study. These properties make this treatment a very attractive method to enhance the performance of polymer coatings.

The increase in impedance with treatment time (figure 4.1.1.6) is related to a increasing amount of formed layer. Due to different factors the treatment with 7 mol L^{-1} HF and 28 mol L^{-1} HF resulted in a lower corrosion protection. The reason for the poor behaviour of the samples treated with of 7 mol L^{-1} HF is the higher hydroxide concentration, as detected by FT-IR spectroscopy. High hydroxide concentrations have a weakening effect on the protective properties of the layer, since that $Mg(OH)_2$ is unstable in the presence of Cl^- ions in solutions. In case of 28 mol L^{-1} HF, the sample was heavily etched and showed a weight loss even after 5 h of treatment, indicating a very slow deposition of protective layer. This suggests that even after 24 h the surface was not completely covered by a layer, resulting in a lower corrosion protection. With a lower quantity of hydroxide and a faster protective layer formation process, the solutions 14 and 20 mol L^{-1} HF resulted in a better corrosion protection.

To conclude this discussion is important to mention that, despite the good properties of the HF-treated samples, this treatment is not suitable for industrial applications due to problems associated to high scale manipulation of HF. The aim for this study was to determine the best conditions for this interesting treatment and investigate its influence on the corrosion protection, as well as on the performance of an upper coating. An interesting outcome of this study is the good performance of coated samples which has interfacial reactions/interactions which results in long term stability of the sample. This concept can be further investigated using more environmentally friendly pre-treatments for the development of industrially viable corrosion protection processes for magnesium alloys

5.1.2 – Acetic and nitric acid cleaning

While HF is a highly dangerous acid and needs to act on the alloy for 24 h to produce its best effect, acetic and nitric acid are much less harmful and can produce good results even after 2 minutes of treatment. This makes these acids interesting candidates for industrial application in the cleaning of magnesium alloys. As reported by Nwaogu [1.33, 1.34], neither acetic nor nitric acid create layers on the substrate surface and the improvement in corrosion rate can be mainly attributed to the removal of impurities. An increase in the quantity of oxides/hydroxides as well as the presence of signals related to nitrates and acetates was observed in the infrared spectra. The presence of these salts at the interface certainly has influence in the osmotic pressure of the system, but however, the dominant parameter related to the lower performance of acid cleaned samples (compared to the ground substrate) is related to the substrate surface roughness.

The roughening of surfaces is a well known procedure in metal and paint industries, which is used to improve the adhesion between a metal and a paint. The relation between surface roughness and adhesion lays on the theory that the adhesion is related to an anchorage effect, which would increase with the number of anchoring spots that are generated by a surface roughening [1.95]. Methods as dry abrasive blasting, water jetting and ultra-high pressure abrasive blasting are well known in the surface preparation of steel substrates as described by Momber [5.3, 5.4]. Despite the assumption that surface roughening increases the adhesion, in the present work no relation between the substrate preparation process and the coating adhesion was observed. Similar results are also reported in studies in the literature [1.37].

An important observation made in the present study is that the coating requires a minimum thickness to avoid defect formation induced by surface roughness. This is especially true when very thin coatings are applied, as in the case of spin-coating. For all three polymers it was observed that the acetic acid cleaned substrate did not performed well neither in EIS nor in immersion tests due to the highly irregular substrate surface. Even when coatings with thickness around 15 µm were applied the performance of the coated acetic acid cleaned substrate was much inferior to that of the other samples. On the other hand, the nitric acid treated substrate showed very good performance, and in case of PVDF and PAN coatings, the performance of this substrate was only inferior to that of the HF-treated one. This difference is related to the much smoother surface produced by the nitric acid cleaning process.

These results indicate that, among the pre-treatments investigated in this study, the cleaning with nitric acid is the more interesting pre-treatment for industrial applications. It is relatively fast and allows the preparation of thin coatings which provide corrosion protection for at least 3 days of immersion in the corrosive solution. Nevertheless the coating should not be as thin as the ones prepared by spin-coating. For such coatings only the grinding pre-treatment leads to acceptable results. Further discussions on the influence of the substrate pre-treatments on the performance of specific coating will follow in the next sections.

5.2 – Poly(ether imide) coatings

5.2.1- Influence of solvent

The two main factors related to the performance of PEI coatings are the type of solvent and the substrate preparation method. The first one has considerably effect in the diffusion of water through the coating as shown by the EIS simulation results due to the retention of residual solvent. The interaction of PEI with DMAc and NMP is so strong that even after 12 h in a vacuum oven (10 mbar) at 130 °C residual solvent was observed. This is one drawback of these solvents compared to methylene chloride, another solvent investigated by our group for the preparation of PEI coatings [1.68]. However, DMAc and NMP have the advantage of allowing the preparation of more concentrated solutions that decreases the solvent waste. Besides that, the solution concentration is important for the final coating thickness. Spin-coating tests were also performed using methylene chloride solutions and unsatisfactory properties were obtained.

Another solvent tested in this work was N,N'-dimethylformamide (DMF) a solvent that differs from DMAc by one single CH_3 group and has a lower boiling point (150 °C) what makes it easier to be removed by heating. However, PEI solutions are unstable in this solvent at concentrations higher than 10 wt.-% and require heating to become homogeneous. Therefore, the solvents DMAc and NMP were selected as the appropriate ones for this application despite the difficulty to completely remove them from the coatings.

Due to the presence of residual solvent a different electronic circuit was required for the fitting process of spin-coated PAN. The coatings prepared by the dip-coating method could be simulated by a tradition circuit, even having high residual solvent amount. This is probably related to a more homogenous solvent dispersion in the coating as its thickness increases. With a more homogeneous solvent distribution there will not be specific solvent-rich and solvent-poor domains and the EIS spectrum can be simulated using the traditional circuits with low error (below 10%). In the case of coatings prepared by the dip-coating method on HF-treated substrates, a similar behaviour to the spin-coated samples is observed in the theta Bode plot, as shown in figure 5.2.1.

It was previously discussed that the capacitance of the coatings on HF-treated substrate was higher than that on ground ones, suggesting that the HF-treated surface could interact with the solvents. This would lead to a higher residual solvent amount in these coatings that could enhance the formation of distinct solvent-rich and solvent-poor domains. By comparing the theta Bode plot of PEI/DMAc coatings on ground and HF-treated substrates it becomes clear the presence of solvent-rich domains in the last one (figure 5.2.1a)

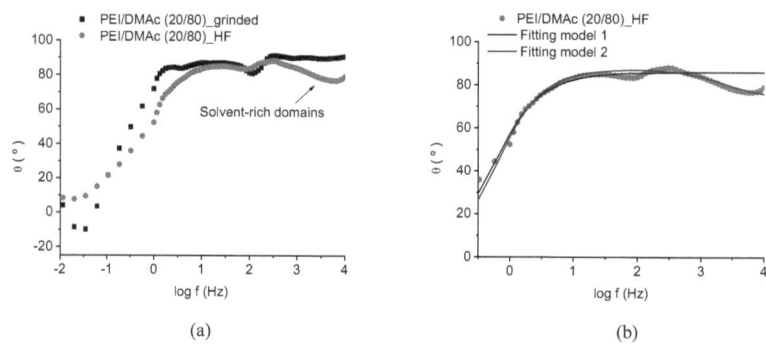

Figure 5.2.1: Theta Bode plots of PEI/DMAc (20/80) coatings on ground and on HF treated substrates.

It can be observed in figure 5.2.1b the difference in fitting the results with the traditional and the circuit considering solvent-rich and solvent-poor domains. The fitting with the traditional circuit results in a straight line at high frequencies without following the spectrum curve, while the fitting using the circuit considering the solvent-rich and solvent-poor domains follows the behaviour of the experimental curve.

However, the influence of different domains in this coating was not so pronounced in the final capacitance as can be observed by comparing figure 5.2.2 (which shows the capacitance variation of the solvent-poor domain of coated HF-treated substrate) to figure 4.2.2.9. It can be observed that the solvent-poor domain capacitance follows the same trend as the coating capacitance calculated using the traditional circuit, and the value difference is of about 0.06 nF.cm^2 corresponding to an error of only 10%. In the case of spin-coated coatings, the error obtained using the traditional circuit was higher than 100% for some spectra. Besides that, after 1100 h of exposure the spectra of dip-coated HF-treated substrates could not be simulated using the circuit considering different domains, suggesting that a part of the solvent was removed and the coating were homogenous. Therefore, the behaviour of this coating could be simulated using the traditional circuits with good accuracy.

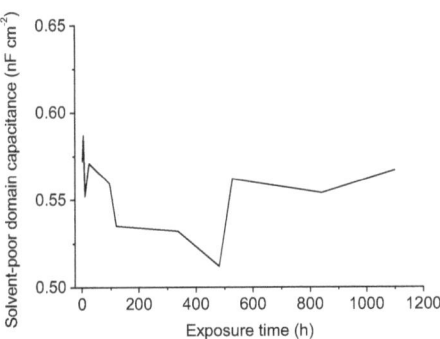

Figure 5.2.2: Capacitance variation of the solvent-poor domain of a coating on HF-treated substrates.

These results suggest that there is a critical solvent amount for a specific polymer volume that will define the presence of different domains in the film or a uniform solvent distribution. Above this critical value, the water diffusion will considerably increase, as observed for just spin-coated PEI. For the present coatings, the critical solvent amount is in the range of 5-6 % for PEI coatings prepared using DMAc solutions, depending on the coating thickness.

The coatings prepared by NMP had higher residual solvent amount and inferior barrier properties compared to the ones prepared by DMAc. These solvents have very different chemical structures, polarities and interact with the polymer with different intensities. It was previously commented that the solvent presence causes some distortion in the polymer secondary structure and that these distortions have influence in the water diffusion. The influence of such distortions on the ether linkage on the diffusion of gases was carefully studied by Kostina et al.[4.23] who performed simulation tests with different angles in the ether segment and showed differences in gases diffusions. Kostina concluded that the diffusion increases as the chain rigidity increases [4.23]. A similar trend is observed in the present study. NMP has stronger interactions with PEI than DMAc producing a higher increase in the rotational barrier, and the diffusion of water is faster for the NMP coatings.

To conclude this discussion about the solvent effects, it is necessary to discuss how the polymer interacts with the solvent. It was previously shown that both solvents interact at the imide ring, due to changes in the carbonyl signals (1700 to 1800 cm^{-1}) and changes in the C-N-C signal (around 1350 cm^{-1}) of the infrared spectrum. However, studies in the literature show interactions of the same and other solvents at other segments of the PEI chain, as in the oxygen in the ether linkage and even the π-electron cloud of the aromatic rings. For example,

Kostina shows that the hydrogen of methylene chloride interacts with the oxygen in the ether linkage of PEI [4.23]. Hatton and Richards [5.5] describe complexes of DMAc and aromatic compounds where they observed interactions between the nitrogen in the solvent with the aromatic ring.

In the present study there is no doubt about an interaction in the imide ring, due to significant characteristics in the carbonyl and on the C-N-C bond signals, as shown in figures 4.2.1.4 and 4.2.2.2. Comparing the spectrum of pure PEI with the one with 4% of residual solvent (figure 4.2.3), changes can be observed in these aforementioned signals. A third carbonyl signal appears, and the signal related to the asymmetrical stretching of the carbonyl group is slightly shifted to higher wave numbers, as can be observed by the dashed line. As mentioned earlier, this shift and decrease in intensity suggests a distortion in the polymer secondary structure, due to a non-co planarity of the two imide rings. It can also be observed that the signal around 1350 cm^{-1} modifies while the signals related to the aromatic ring (C=C) and the ether linkage (Ph-O-Ph) are unchanged at this residual solvent content. This clearly shows that the solvent interacts with the imide ring.

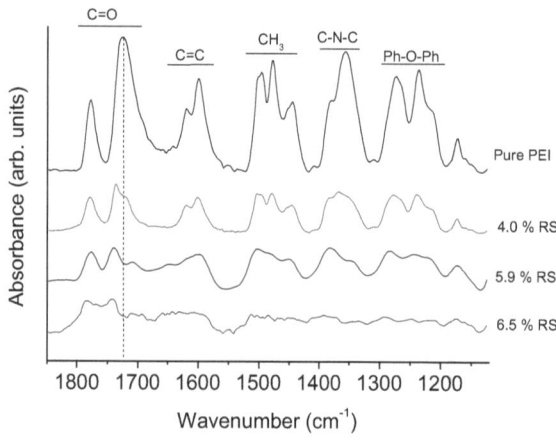

Figure 5.2.3: FT-IR spectra of PEI coatings prepared using DMAc solutions with different amount of residual solvent (RS).

As the solvent content increases, significant changes happen to the other signals. In fact, the entire spectra changes with new signals appearing and variations in signal intensities. This result shows that the primary interaction of DMAc with PEI is at the imide ring, but as the residual solvent content increases, it is possible that interactions at the aromatic rings and at the ether linkage also appear. Similar analyses were performed on coatings prepared using NMP and the same conclusions were obtained.

5.2.2 Influence of substrate pre-treatment

The substrate pre-treatment has considerable influence in the performance of PEI coatings. Three parameters of importance were observed: the substrate surface roughness, substrate impurity level and the formation of compounds at the substrate surface. The first one has influence on the morphology of the coating and can form defects which considerably decreases the protectiveness of the coatings. The second influences the corrosion rate at the interface and the last influences the interfacial stability and adhesion of the coating. The effect of these three parameters in the performance of PEI coatings will be discussed in the following.

It was observed that the negative influence of high substrate surface roughness decrease as the coating thickness increases. Aiming to have better insights on this relation, the ratio between these two parameters was measured for different samples and this ratio was correlated to the sample impedance after 72 h of exposure to 3.5wt-% NaCl. The impedance at this time was selected as a reference point to give a better insight on the long term performance of the coatings. The results of this investigation are shown in table 5.2.1

Table 5.2.1 Coating thickness and substrate roughness influence on the impedance.

Solution/substrate	Coating thickness (μm)	Substrate roughness (μm)	CT/SR*	Impedance at 72h ($M\Omega\ cm^2$)
PEI-DMAc/ground	2.5 μm	0.09	28	12.4
PEI-DMAc/ground	5 μm	0.09	55	25.8
PEI-DMAc/ground	13 μm	0.09	144	2500
PEI-DMAc/AR	5 μm	0.37	13	degraded
PEI-DMAc/AR	8 μm	0.37	21	degraded
PEI-DMAc/AR	13 μm	0.37	35	221
PEI-DMAc/AA	2.5 μm	2.12	1.8	degraded
PEI-DMAc/AA	5 μm	2.12	2	degraded
PEI-DMAc/AA	13 μm	2.12	6	degraded
PEI-DMAc/AN	2.5 μm	0.36	7	degraded
PEI-DMAc/AN	-	-	-	-
PEI-DMAc/AN	13 μm	0.36	36	16.5
PEI-DMAc/HF	2.5 μm	0.37	7	0.3
PEI-DMAc/HF	8 μm	0.37	21	17
PEI-DMAc/HF	13 μm	0.37	35	1523

*CT: Coating thickness; SR: substrate surface roughness

For the ground, acetic and nitric acid cleaned substrates, the higher the CT/SR (coating thickness/surface roughness) the higher the coating impedance at 72 h of exposure. The results also show that the CT/SR ration should have a minimal value of 21 to maintain good impedance after 72 h of exposure. In case of special interfacial interaction (as in the case of coated HF-treated substrates) this value may decrease since that the interface interaction plays the most significant role in the sample performance. The as-received coatings follow a similar trend observed for the ground, acetic and nitric acid cleaned samples. However, it can be observed that with a CT/SR value of 35 this sample has an impedance of 221 $M\Omega\ cm^2$ while the substrate treated with nitric acid has lower impedance with slightly higher ratio. This is probably associated with an interaction between the oxide layer in the metal surface and the polymer.

Based on these results it can be concluded that, in the lack of beneficial interfacial interactions, a PEI coating will only produce satisfactory protection when the CT/SR ration is higher than 21. This can be achieved either by treating the substrate in a way that its surface roughness decreases or increasing the coating thickness. The only substrate treatment that satisfies this condition for spin-coating is the grinding. As the grinding process is not suitable for industrial applications different pre-treatments should be investigated in order to apply the spin-coating of PEI in industrial processes for corrosion protection of magnesium alloys.

Besides the effect on surface roughness, the different used pre-treatments have considerable effect in the impurity removal. All the three acids used in the present study have an efficient impurity removal rate as previously commented in the section about HF treatment and in the works of Nwaogu [1.33, 1.34]. In some cases the surface impurity concentration does not play a significant role in the coating performance. For example, the performance of as-received coated substrate was superior to that of the acetic acid cleaned coated one. It is certain that the corrosion rate of the as-received coated substrate will be higher than that of the clean one when water concentrates at the interface. However, the water diffusion rate and its capacity to concentrate in the interface is the most significant parameter for the general sample performance.

The importance of clean surfaces can be observed by comparing the result of nitric acid cleaned samples with the as-received one. The roughness of these substrates is very similar (table 4.1.2.1) but the performance of the clean one is much better. The as-received dip-coated sample showed high impedances only during the first 3 days of exposure while the nitric acid cleaned maintaining impedance in the order of 10^7 Ω cm^2 for 7 days (see figures 4.2.2.5 and 4.2.2.7). A proper surface cleaning can then double the long term corrosion stability of a sample. As all the used acids have similar effects on the impurity removal, other parameters are more relevant in the selection of the most adequate one for a particular application. The CT/SR ratio, the presence of strong interfacial interactions and the easiness of the treatment should lead to the appropriate choice.

As extensively discussed in the results section, the best performance of the coating in HF-treated substrate is related to an interfacial interaction. As the as-received substrate surface contains hydroxides and oxides which could also interact with the polymer, the interface of this system was also investigated. The XPS spectra in figure 5.2.4a shows that the binding energy of Mg 2p electrons did not shift to higher BE values in case of coated as-received substrates as it does in case of HF-treated ones. This is probably related to a lower basicity of the oxygen atoms as observed in figure 5.2.4b. It can be observed that the binding energy of the oxygen electron has lower values on the HF-treated sample than in the as-received one. This lower binding energy values indicate a higher electron density, and consequently, higher basicity. This result is in accordance with the study of Prescott et al. [4.13] that shows that the oxygen basicity in the O/F mixed lattice is higher than in the single one.

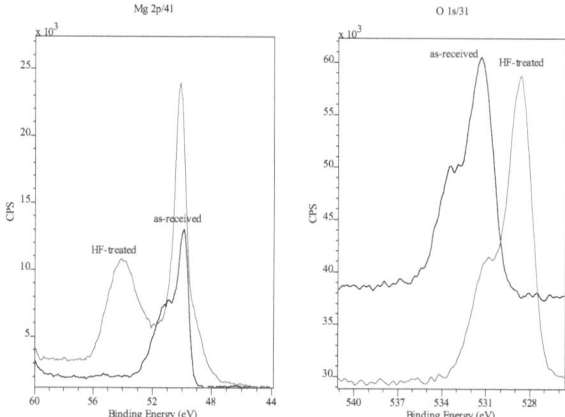

Figure 5.2.4: XPS spectra of magnesium and oxygen electrons at the interface of as-received and HF-treated coated samples.

It is interesting to mention that some papers describe that the imide ring is electron deficient and forms complexes with electron donating specimens as metals like chromium and potassium [5.6, 5.7]. Considering this, the interface in ground substrate was investigated by XPS spectroscopy to proof if such complexes were formed. No shifts confirming this hypothesis were observed. Theses mentioned studies are related to polyimides containing the pyromellitic dianhydride (PMDA) moiety, which has one aromatic ring between two imide rings, as shown in figure 5.2.5. For the structure of PMDA it is easy to realize that the aromatic ring between the imide rings is electron deficient since that the four carbonyl groups linked to it have an intense electron withdrawing effect. In the case of PEI Ultem there is also an aromatic ring between two imide rings, but the linkage is on the nitrogen atom rather than in the carbonyl groups (see figure 4.2.2.2). The nitrogen atom withdraw electron by inductive effect (more electronegative) but donate electron by mesomeric effect (delocalization of the π-cloud). As for nitrogen the mesomeric effect is more pronounced it is expected that the density of this ring will actually increase instead of decrease. Therefore, in the case of Ultem there is no aromatic ring with electron deficiency similar to the one in the PMDA structure to form such complexes.

126

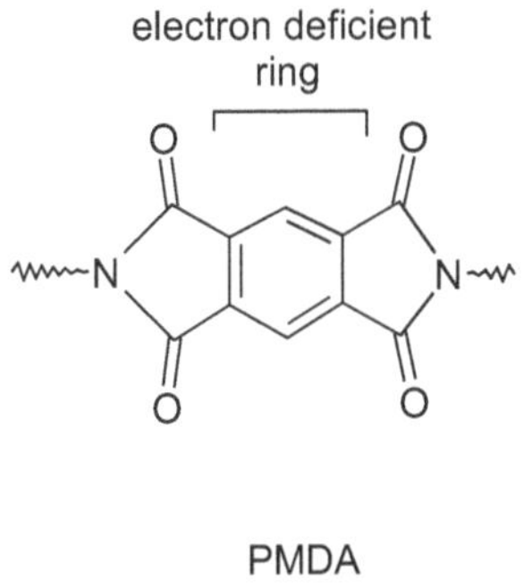

electron deficient ring

PMDA

Figure 5.2.5: Structure of the polyimide PMDA.

5.2.3 – Mechanism of coating degradation: Interfacial reactions

To investigate whether any reaction between the polymer and the corrosion products took place during the corrosion of the samples (e.g. opening of the imide ring as postulated by Scharnagl et al. [1.68] as shown in figure 5.2.6) IR microscopy analyses were performed in spin-coated samples after exposure. This investigation is important for a better understanding of the mechanism of coating degradation. Figure 5.2.7a shows the microscopic view of a sample spin-coated with PEI/DMAc (10/90) after the corrosion test, where an undamaged and damaged area can be seen. The IR spectra of these areas, figure 5.2.7b, reveal the presence of O-H signals on the damaged area, which are related to magnesium hydroxide. The sharp signal at 3700 cm^{-1} is related to brucite crystals (Mg(OH)$_2$) which do not form hydrogen bonds [4.10]. The broad signals in the range from 3000 and 3500 cm^{-1} are related to magnesium hydroxide forming hydrogen bonds and probably to polyamic acid [4.9, 4.10].

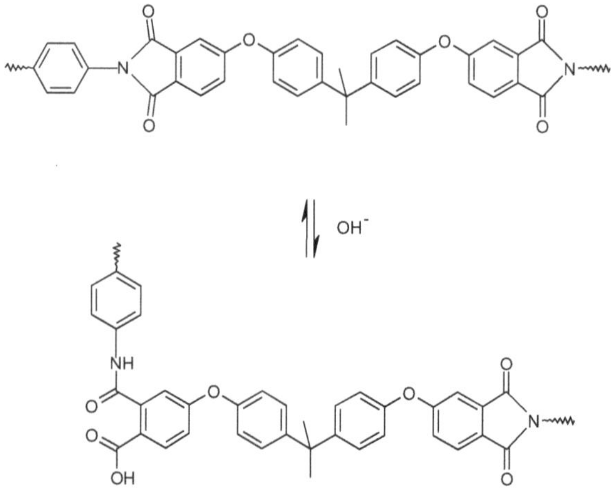

Figure 5.2.6: Scheme of the imide ring opening produced by the polymer reaction with hydroxides anions.

To evaluate whether the ring opening reaction took place, the ratio of the signal related to the carbonyl group and the one related to the ether linkage (Ar-O-Ar) of the PEI structure were investigated for the damaged and two undamaged areas, indicated by the colours blue and red in figure 5.2.7a. The carbonyl/ether ratio considerably decreases from the undamaged to the more damaged area by the following values: 4.91 (undamaged), 3.11 (blue area) 2.56 (red area). This result clearly shows that the quantity of imide rings decreases in the polymer structure indicating the reaction at the interface. Besides that, new signals appeared in the spectrum of the more damaged area that can be attributed to the polyamic acid and to magnesium polyamate (salt form of polyamic acid) as follow: 3500-300 cm^{-1} (O-H of polyamic acid, water and $Mg(OH)_2$ with hydrogen bonds), 1660 cm^{-1} (amide I), 1560 cm^{-1} (amide II), 1522 cm^{-1} (aromatic ring of polyamic acid) and 1400-1390 cm^{-1} (carboxylate anion) [1.85, 1.87].

(a)

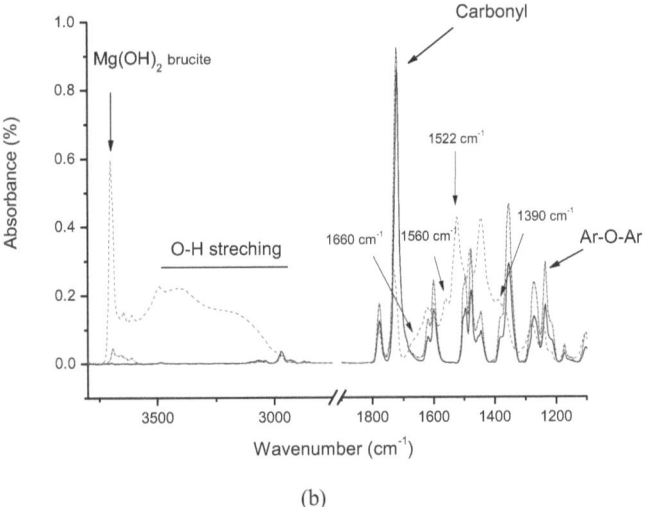

(b)

Figure 5.2.7: (a) Microscopic image of a ground substrate spin-coated with PEI/DMAc (10/90) under N_2 atmosphere after 24 h of exposure to 3.5 wt.-% NaCl solution. (b) IR spectra of three distinct points of the sample surface.

The presence of polyamate can further be confirmed by XPS spectroscopy, as shown in figure 5.2.8. It can be observed that the binding energy of the magnesium 2p electrons is shifted to higher values in the damaged coating compared to the signals of usual magnesium corrosion products (magnesium oxide and magnesium hydroxide). This signal is probably related to the Mg^{2+} cation which stabilizes the polyamate anion. Therefore, it can be affirmed that magnesium hydroxide reacts with the imide ring forming magnesium polyamate and polyamic acid.

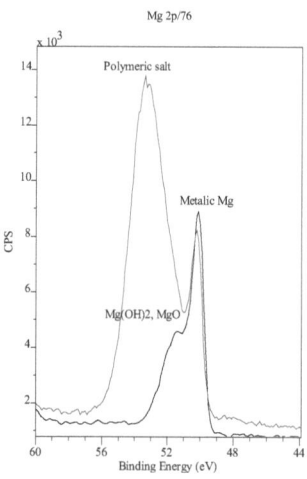

Figure 5.2.8: XPS spectra of magnesium electrons at the interface of a damaged PEI coating on ground substrates.

The effect of this interfacial reaction in the protectiveness of the coating was investigated by analyzing the Nyquist plot of the post-dried spin-coated sample, shown in figure 5.2.9. It can be observed that at the exposure time of 288 h there are two distinct semicircles with similar magnitudes. These two semicircles relate to the coating capacitance and to interfacial capacitance (double layer capacitance) [1.95]. It is interesting to observe how the intensity and proportion of these semicircles changes with exposure time. It can be seen that from 288 h to 336 h the first semicircle (related to the coating) do not change while the second one (interface) considerably increases, becoming bigger than the first one. This increase is probably related to the deposition of magnesium hydroxide in the interface, which has low solubility in water and would partially protect the metal surface.

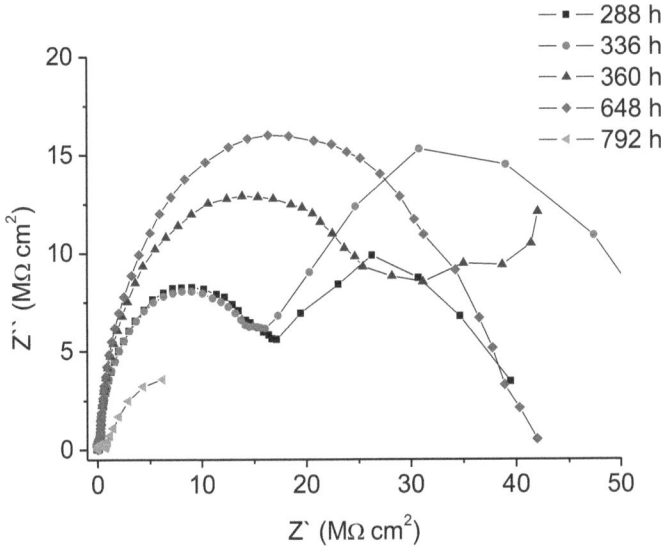

Figure 5.2.9: Nyquist plot of a spin-coated sample dried in a vacuum oven at different exposure times.

However, from 336 h to 648 h the opposite happens. At this time, the coating impedance considerably increases and the second semicircle is eliminated. This can be explained by the reaction of magnesium hydroxide with the polymer that would disturb the dissociation equilibrium of magnesium hydroxide in the products direction decreasing the amount of deposited base. The increase in coating resistance suggests that this reaction did not happen exclusively at the interface but also in the bulk coating (what is a very reasonable hypothesis since that magnesium hydroxide dissolved in water could diffuse through the coating) and that the formed polyamate has higher resistance than the pure PEI.

From 648 h to 792 h there is a drastic decrease in impedance and 24 h after that the impedance was in the same order of the uncoated metal. This probably relates to the transformation of polyamate to polyamic acid by the salt contact with water, which has higher water affinity and would enhance the water diffusion through the coating, decreasing its resistance. Considering the proposed mechanism, it can be concluded that the formation of magnesium polyamate increases the impedance of the coating and helps to maintain this high impedance for at least 300 h. However, when magnesium polyamate is converted to polyamic acid, the coating impedance suffers a considerable decrease due to the higher water affinity of the polyamic acid unit. In this sense, the formation of magnesium polyamate has a positive effect in the protection of the sample while the formation of polyamic acid has a negative one.

This result suggests that this sample performance could be improved if the conversion of magnesium polyamate to polyamic acid could be inhibited or if the formed polyamic acid could increase the interfacial stability of the system by interacting with the substrate.

This second option seems to take place in the case of PEI coatings on HF-treated substrates. In the case of HF-treated dip-coated substrate, a similar increase in the interfacial impedance can be observed in the bode plot from the exposure time of 2540 h to the time of 3312 h, as shown in figure 4.2.2.7b. The presence of polyamate could also be confirmed in the XPS spectra of the interface of this damaged sample as shown in figure 5.2.10. It is shown that the polyamate is formed only in the interface, and the HF-treated substrate maintains the same signals at higher depths (the signal above 57 eV is related magnesium fluoride interacting with the coating and the one at 51.5 eV to magnesium hydroxide). As previously commented, the acid-base interaction will become stronger as the magnesium polyamate is converted to polyamic acid as it has a much stronger acid character.

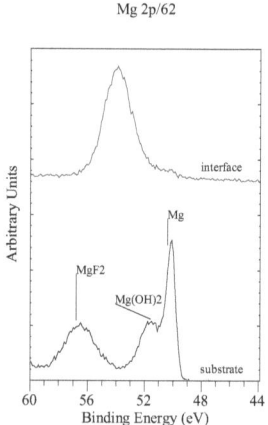

Figure 5.2.10: XPS spectra of the interface of a PEI coating on HF-treated substrate after exposure to the corrosive solution.

Therefore, besides the initial stronger interfacial stability of the HF-treated coated substrate, the best performance of this system is also associated to the substrate interaction with the formed polyamic acid. This can be the reason for the drastic impedance decrease that takes place for the coated ground substrate for the exposure time of 1730 h to 1992 h as shown in figure 4.2.2.3a while the HF-treated coated substrate shows impedance increase in a similar exposure time.

5.3 – PVDF coatings

5.3.1 – Influence of solvent

Different from PEI, PVDF is a semi-crystalline polymer where the crystallinity degree and type of crystalline phase can change with the processing of the polymer. Therefore it could result in coatings with distinct performances in corrosion tests depending on how they were prepared. It can be seen from table 5.3.1 that neither the crystallinity nor the melting temperature (crystalline phase) significantly changes with solvents and drying temperature. The main crystalline phase is α in all conditions (β and γ have melting points above 180 °C) [4.27] that is not a unexpected result since that it is the most common crystalline phase of PVDF. The literature [4.25, 5.9] reports that considerable amount of the other phases can only be formed by specific processes. Some of these processes are the casting of much diluted solutions at specific drying temperature and cooling the melt at certain conditions. An interesting method described in the literature for the preparation of PVDF films mainly in the β phase is to apply a specific stress in α-phased films, as reported by Lando [5.8].

The crystallinity slightly decreases when the polymer is dried at 180 °C indicating that the used cooling rate (the polymer was removed from the oven and place at room atmosphere) was considerably high to not allow the crystallization of all melted crystals. The samples dried at the same temperature showed similar crystallinity degree and melting temperature, with the exception of. the PVDF/DMAc (20/80) coating dried at 180 °C . This sample showed the lowest crystallinity degree. Two analyses were performed for this sample and the same result was obtained. No apparent reason for this result could be found.

As the crystallinity slightly decrease when the coatings were dried at 180 °C the increase in the number of pinholes shown in figure 4.2.4.2e is not related to a crystallinity increase as previously mentioned, but rather to changes in morphology produced by the melt-recrystallization process. The lack of dependence of the crystalline phase to the used solvent is probably a result of the weak interactions between the polymer and solvents. As the interactions are only strong enough to dissolve the polymer, different orientations are not likely to occur by the use of different solvents. Besides that, no residual solvent was observed in any of the prepared coatings. Due to the similarities in morphology, crystalline phase, degree of crystallinity and residual solvent amount in the coatings prepared with different solutions, the performance of one coating is representative for all others dried at the same temperature. The one prepared using DMAc solutions was selected for further characterizations to have a better comparison with PEI. As the different solvents do not produce coatings with different properties and no residual solvent was observed, the choice of the appropriated solvent for a particular application can be made based on other parameters, as for example, economic factors.

Table 5.3.1 Crystallinity degree of the coatings prepared in different conditions.

Solution	Crystallinity (%)	Melting temperature (°C)
PVDF/DMF (150 °C)	74	169
PVDF/DMF (180 °C)	71	168
PVDF/NMP (150 °C)	70	167
PVDF/NMP (180 °C)	68	169
PVDF/DMAc (150 °C)	73	167
PVDF/DMAc(180 °C)	61	167

5.3.2 – Effect of substrate pre-treatment

The substrate pre-treatment has a major influence in the protective properties of PVDF coatings, even higher as for PEI. It was observed that, while the performance of PVDF coatings on ground substrate was very poor, it was very good on HF-treated substrate. This difference is related to the adhesion of the polymer. PVDF is known as a low adhesion polymer, a property that is related to the low polarizability of the C-F bond, and usually requires specific substrate pre-treatments, or chemical modifications in the polymer structure, to be industrially applied as coating [1.81-1.83]. While on the HF-treated substrate, the interfacial interaction provided sufficient stability for the coating, on ground substrates the lack of interactions resulted in delamination after a few minutes of immersion.

The best performance was observed for the HF-treated substrate. PVDF coatings on nitric acid cleaned substrate also showed good performance. The adhesion on this substrate was slightly superior to that on ground one as shown in table 4.2.4.3. This is related to the incomplete removal of magnesium oxide/hydroxide (observed by infrared analyses) during the nitric acid treatment. These compounds also interact with the polymer, as observed for the higher coating adhesion on as-received substrate. The as-received substrate had a superior performance than the ground one, due to this better adhesion which probably arises from interactions between the basic substrate surface and the acid methylene hydrogen on the polymer structure. However, the as-received coating substrate was not stable for more than 2 days in the corrosive solution due to surface impurities, while the nitric acid cleaned one was stable for three days. The performance of acetic acid cleaned samples was similar to the observed for ground substrates. This shows that the substrate surface roughness does not play a significant role for the coatings adhesion. At the evaluated coating thickness, no influence of CT/SR was observed for PVDF coatings.

This results show that chemical interaction is a much more significant parameter for PVDF coatings adhesion to AZ31 sheets than mechanical interlocking. Therefore, any treatment that aims to improve the adhesion of these coatings must increase the polarity of the substrate surface or the polarity of the polymer structure without necessarily increasing the roughness. A magnesium oxide/hydroxide layer on the substrate surface slightly increases the coating adhesion. However, the adhesion increase rendered by such layer is still low, especially when compared to the adhesion of PEI coatings (1.37 MPa while the adhesion of PEI is 5.38 MPa). The adhesion improved rendered by the HF treatment is the highest among the tested pre-treatments (2.22 MPa). The adhesion strength of PVDF coatings on HF-treated substrate is still much inferior to that of PEI coatings, but enough to render good performance, as observed in impedance and immersion tests.

5.3.3 – Mechanism of coating degradation

It was demonstrated in section 4.2.4.2 that the coated HF-treated substrate shows a constant value of coating resistance and capacitance after some time of exposure, suggesting the occurrence of interfacial processes that brings the system to an equilibrium. One possible process responsible for this behaviour is the occurrence of interfacial reactions between the polymer and the corrosion product magnesium hydroxide, similar to the observed for PEI coatings. It is well known in the literature that PVDF can react with bases to eliminate fluoride and form a double bond, as schematically shown in the step I of figure 5.3.1 [1.82, 5.10, 5.11].

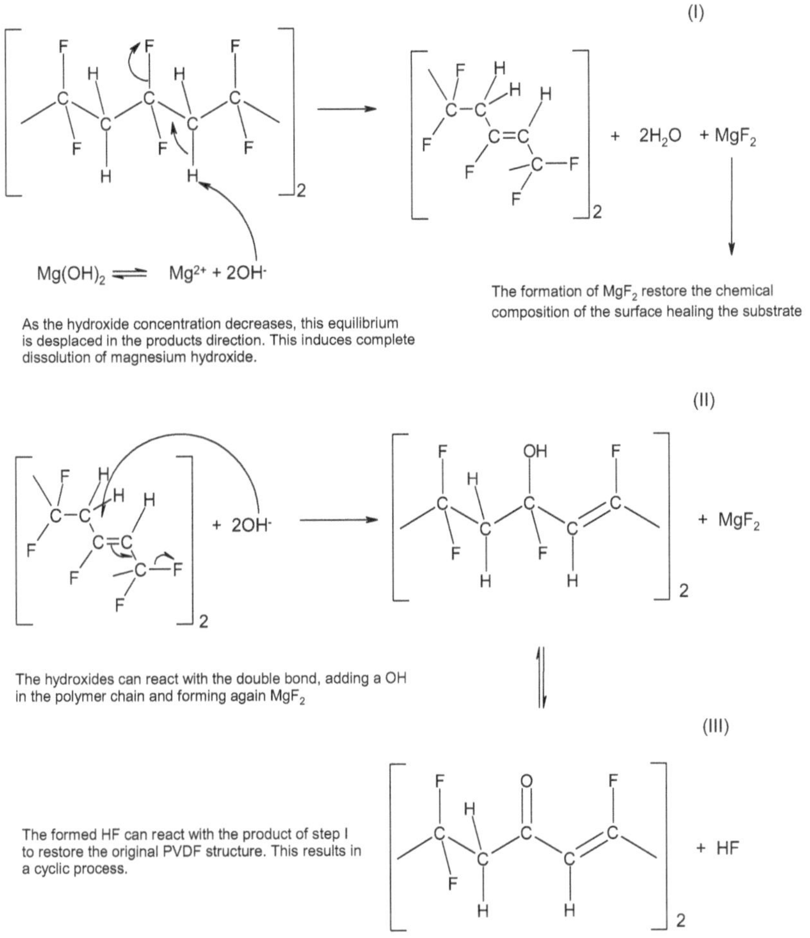

(I)

Mg(OH)$_2$ ⇌ Mg^{2+} + 2OH$^-$

As the hydroxide concentration decreases, this equilibrium is desplaced in the products direction. This induces complete dissolution of magnesium hydroxide.

The formation of MgF$_2$ restore the chemical composition of the surface healing the substrate

(II)

The hydroxides can react with the double bond, adding a OH in the polymer chain and forming again MgF$_2$

(III)

The formed HF can react with the product of step I to restore the original PVDF structure. This results in a cyclic process.

Figure 5.3.1: General scheme of the interfacial processes.

As hydroxide is formed by the arrival of water at the interface it is possible that this elimination reaction takes place. This would increase the formation of magnesium fluoride in the substrate surface (see figure 5.3.1). In this case, a self-healing process would happen with the conversion of the less protective magnesium hydroxide, formed during the corrosion process, in the more protective magnesium fluoride, restoring the initial condition of the substrate surface. For this mechanism all the formed Mg(OH)$_2$ would be completely converted to magnesium fluoride since that its reaction with the polymer would disturb its dissociation equilibrium in the products direction.

To confirm the occurrence of this reaction the interfacial side of a PVDF coating on HF-treated substrate after 3 months of exposure to 3.5 wt.-% NaCl was examined using XPS analyses. For this analysis, the coating was removed from the substrate using a sharp blade instead of using sputtering to avoid degradation of the polymer [5.11]. The spectra, shown in figure 5.3.2, clearly show the occurrence of interfacial reactions which changed the chemical composition of the film. It can clearly be seen that the intensity of the CF_2 signal decreases in relation to the CH_2 one compared to the surface side of the coating. In table 5.3.2 it can be observed that the atomic concentration of fluoride decreased from 43% to 33% confirming the occurrence of defluorination.

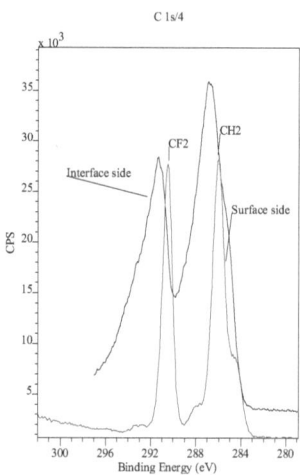

Figure 5.3.2: XPS spectra of the surface of an undamaged PVDF coating and of the interface side of a coating after 3 months of exposure to the corrosive solution.

It is interesting to observe in table 5.3.2 that the atomic concentration of fluoride decreases in the same quantity as the concentration of oxygen increases indicating the addition of oxygen in the polymer chain after fluoride removal. Besides that, it can be clearly seen in the C 1s XPS spectra that the width of the two peaks considerably increases indicating the formation of other carbon bonds besides the double bond (oxygenated derivatives would appears between 285 eV and 292 eV). The increase in these peak widths can be attributed to the formation of hydroxyl and carbonyl functionalities as described by Brewis et. al[5.9] and Ross et. al [5.10]. In the O 1s spectra, a strong signal at 530 eV (related to carbonyl oxygen) and a weak one at 535 eV (related to hydroxyl oxygen) were observed confirming the formation of these functionalities. This results show that the interfacial reaction do not stop on the defluorination step, but rather moves forward in the formation of oxygenated derivatives. A complete mechanism of the reactions at the interface is schematically represented in figure 5.3.1. It is important to mention that as the concentration of fluoride decreases at the same proportion as the concentration of oxygen increases, and that the formation of one mol of oxygenated product is formed by the elimination of, at least, two mol of fluoride it can be assumed that only a part of the formed double bonds react further to form oxygenated derivatives.

Table 5.3.2: XPS results on the atomic concentration at the surface and interface side of a coating.

Sample	Fluoride (atomic %)	Oxygen (atomic %)
Surface side	43.51	1.88
Interface side	33.35	12.71

It is important to comment that beside reacting with hydroxides to form oxygenated functionalities, the double bond formed in step I can react with the HF formed in step III (equilibrium between the FC-OH and C=O) to restore the original PVDF structure. This results in a cyclic process which would continue until the coating is sufficiently adhered to the substrate and until the rate of MgF_2 formation is high enough to convert all the formed $Mg(OH)_2$. Another important aspect that should be mentioned is that only low concentrations of sodium and chloride (approx. 1.4% atomic concentration) could be detected at the interfacial side of the PVDF coating after 2200 h of exposure to 3.5 wt.-% NaCl solution. This shows the good barrier property of the PVDF coatings in relation to ionic specimens. This is an important result for the understanding of the interfacial process. If high concentrations of NaCl could reach the interface it would be necessary to consider its effect on the interfacial process, as for example, the possible formation of NaF. Nevertheless, neither sodium nor chloride could be detected among the corrosion products after the long exposure to the corrosive solution, suggesting that the interfacial process is only governed by the arrival of water at the interface.

Therefore, the constant value of the charge transfer resistance, observed in Figure 4.2.4.3 can be explained considering the conversion of $Mg(OH)_2$ to MgF_2 at the interface while the constant value of the coating resistance can be explained by its reaction with hydroxides which result in compounds with probable lower ionic diffusion coefficients and higher adherence to the substrate. A water back diffusion (from the interface to the coating surface) of ions, water and hydroxides and the fill up of pinholes is responsible for the decrease and constant values of capacitances (figure 4.2.4.4), respectively. However, all these interfacial processes would not be able to render high long-term stability to the sample without good adherence of the coating to the substrate. That is the reason why the performance of PVDF coatings was much superior in HF-treated than in the other substrates.

Besides that, the formation of oxygenated functionalities will have considerable effect on the interfacial stability, since they increase the polymer adhesion [5,10]. The formation of carbonyl and hydroxyl groups will increase the polarity of the film that could enhance the acid-base interaction with the substrate. It is important to mention that certain amount of PVDF remained adhered at the substrate surface after the coating removal for the XPS analyses indicating strong adherence forces. A total concentration of 27 % of carbon could be detected by XPS analyses at the substrate surface which is related to oxygenate derivatives of PVDF and probably to carbonate corrosion products (the oxygen concentration was 19%). To better characterize the compound attached to the substrate, FT-IR microscopy analyses were performed. Figure 5.3.3a shows the microscopic image of this substrate (15 x magnification) where a damaged and an undamaged area (both exposed to the corrosive solution) can be observed. The spectra of the marked points are shown in figure 5.3.3b. For the two points of the undamaged area (blue and black) the signals related to the uncoated substrate (see figures 4.1.1.4 and 4.1.1.5), which should appear below 1000 cm^{-1}, are very weak or absent. On the other hand, it can clearly be identified the presence of carbon compounds by the signals between 1200 and 1600 cm^{-1}. This further corroborates the presence of a thin film of PVDF derivative attached to the substrate even after coating removal.

The broad signal above 3000 cm^{-1}, related to O-H stretching, confirms the presence of oxygenated PVDF derivatives. In fact, the uncoated substrate also has signals at this wave number (see figure 4.1.1.4) but considering the low intensity of the signals below 1000 cm^{-1} it can be concluded that the hydroxyl signals observed in the spectra are related to PVDF derivatives. The signals related to the substrate can be observed in the spectrum of the red spot, which is related to a corroded area.

The lack of a strong signal between 1600 and 1700 cm^{-1} (C=O) suggests that the majority of the oxygenated derivatives attached to the substrate have the hydroxyl group instead of the carbonyl one. On the other hand, on the coating interfacial side, there is a strong carbonyl signal on the O 1s XPS spectrum, as previously commented. This suggests that all the steps shown in figure 5.3.1 happen. Step III is of particular importance for the long term stability of the interfacial process, as it generates HF which can restore the initial condition of the coating.

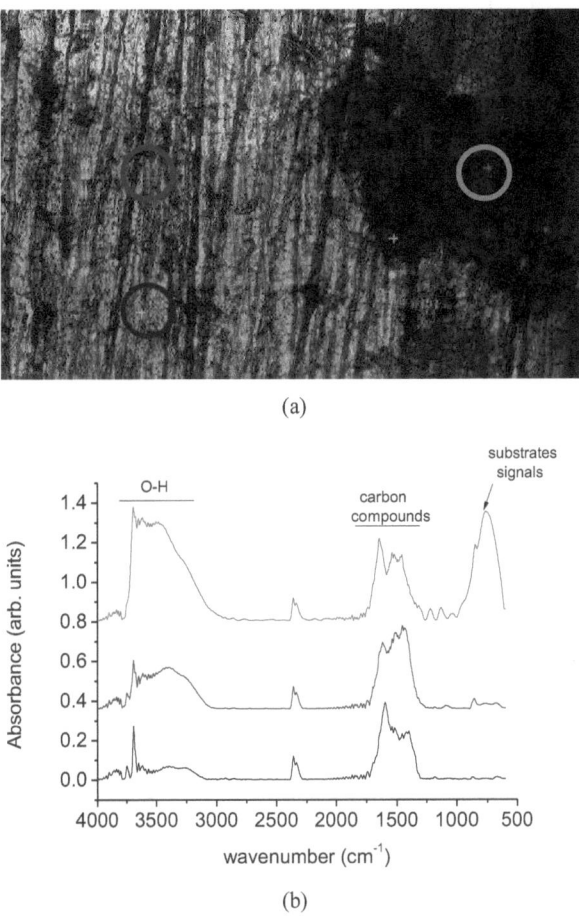

Figure 5.3.3: (a) Microscopic image of the damaged substrate after 3 months of exposure to the corrosive solution. (b) FT-IR spectra of specific points at the substrate surface.

Figure 5.3.4 shows that the film attached to the substrate has a chemical structure very different in comparison to the unmodified polymer. The signals related to the stretching modes of CH_2 and CF_2 are not present, as well as many bending modes in the range of 500 to 1200 cm^{-1} and the stretching mode of carbon-carbon single bond, which should appear at 1150 cm^{-1}. The lack of these signals shows that the film is mainly composed of carbon-carbon double bonds with oxygenated functionalities. Another support for this conclusion is the appearance of a new and strong signal at 1600 cm^{-1} which can be attributed to this carbon double bond stretching.

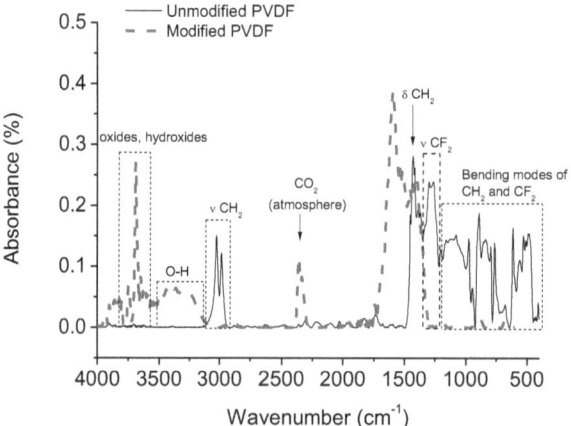

Figure 5.3.4: Infrared spectra of a pure and a modified PVDF film.

The stretching of carbon double bonds usually has very low intensity in infrared spectroscopy due to the lack of changes in the dipole moment (but it is visible in Raman spectroscopy). However, if one of the carbons is bonded to a hydroxyl group a dipole moment appears and the double bond stretching becomes strong. The film that remains attached to the substrate is mainly composed of carbon double bonds with hydroxyl groups, and a possible structure of this is shown in Figure 5.3.5. Some CH_2 are still present as suggest by the presence of a signal at 1430 cm^{-1} related to its δ bend mode.

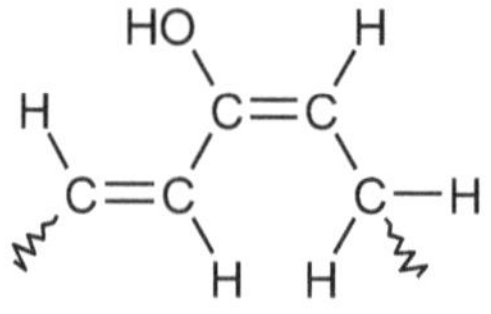

Figure 5.3.5: Possible structure of the film attached to the substrate.

To complete this discussion and emphasize the good performance of this system, an image of the substrate of the sample exposed for three months to the corrosive solution is shown in figure 5.3.6. It can be observed that even after this long exposure time only some small corrosion spots can be visualized at the surface. This shows the high potential of the system PVDF-coating/HF-treatment for the corrosion protection of magnesium alloys. However, similarly to PEI coatings, the long-term stability was much shorter in immersion than in EIS tests due to coating defects at the samples edges (Figure 5.3.7). The higher density of larger pinholes at the edges induces a to high hydroxide formation rate to allow to substrate healing by the interfacial process.

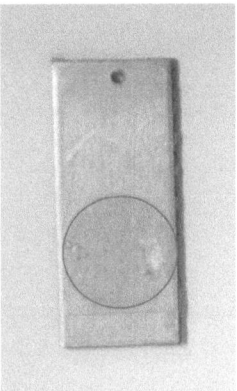

Figure 5.3.6: Image of the substrate exposed for 3 months after the coating was peeled off. The circle indicates the exposed area.

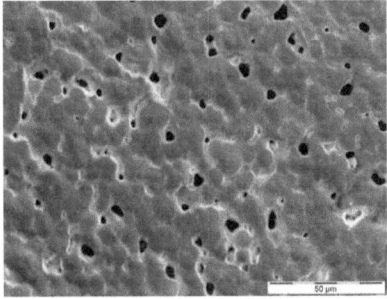

Figure 5.3.7: SEM image of coating prepared using PVDF/DMAc (15/85) solution and dried at 150 °C close to the sample edge.

5.4 – PAN coatings

5.4.1 – Influence of solvent

As observed for PEI coatings, the presence of residual solvent had considerable influence on the coating performance. Table 4.2.5.1 shows that the capacitance of the spin-coated PAN dropped to the half after the drying process and that the resistance increased one order of magnitude. Nevertheless, the influence of residual solvent was not as intense as it is in case of PEI. This can be related to the distribution of the solvent on the film. As previously demonstrated, in case of PEI, the electrolytes start entering in the coating through well defined solvent-rich domains. Such domains were not observed for PAN. The lack of solvent-rich domains in PAN is related to a homogeneous solvent distribution in the coating.

Tests using DMAc and NMP solutions were also performed but disappointing coating performance was observed. In case of PAN in DMF some studies in the literature [5.13, 5.14] discuss the formation of a complex which complicates the total removal of the solvent. The formation of such complex is usually regarded as the reason for the superior properties of PAN components prepared from DMF solutions, in comparison with other solvents [4.38]. This complex is usually considered as an interaction between the nitrile group and the carbonyl carbon, or the nitrogen of the solvent. However, despite the presence of DMF signals in the infrared spectra shown in figure 4.2.6.3 no shifts in the nitrile signal were observed, comparing the PAN powder and the PAN coating, to corroborate this assumption. Nevertheless, the DSC analyses show an interesting influence of the solvent presence on the exothermic peak related to the cyclising process of PAN. It was previously commented that in temperatures close to 300 °C the nitrile group can cyclises as shown in figure 5.4.1, eliminating ammonia or hydrogenated compounds. This process results in a highly exothermic peak in the DSC diagram, as shown in figure 5.4.2. The analysis of this peak showed that for the powder PAN a heat of 393 J g^{-1} is evolved while a value of 553 J g^{-1} and of 508 J g^{-1} is evolved for the just spin-coated and the post-dried coating, respectively. Besides that, there is a shift in 3 °C to lower temperatures in the peak position from the powder to the coating.

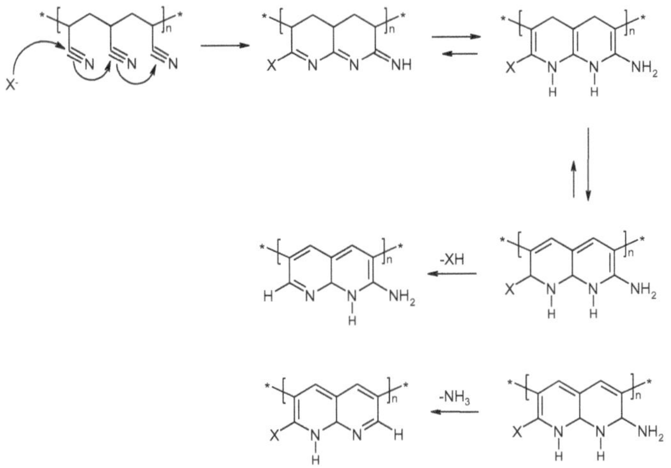

Figure 5.4.1 Scheme of the thermal degradation of PAN. In the scheme, "X" can be impurities or chain defects.

The higher heat evolved in the presence of residual DMF can be explained considering the higher molecular energy of the polymer produced by the plasticizing effect of DMF. The solvent DMF plasticizes the polymer (the T_g of PAN goes from 97 °C to 91 °C with 5 wt.-% of DMF), which means that the polymer chain gets more mobile, in other words, more energy. This additional energy is released during the cyclising process resulting in higher evolved heat. Besides that, the higher energy of the polymer leads to the occurrence of the cyclising process at lower temperatures. Therefore, it can be assumed that certain amount of solvent was still present in the polymer even at the temperature of 280 °C, the onset of the exothermic peak. As the boiling point of this solvent is 150 °C and great part of it is eliminated between 130 – 230 °C as obtained by TGA analyses, it can be assured that strong interactions between the polymer and the solvent takes place. Nevertheless, the infrared spectra indicates that this interaction do not take place at the nitrile group.

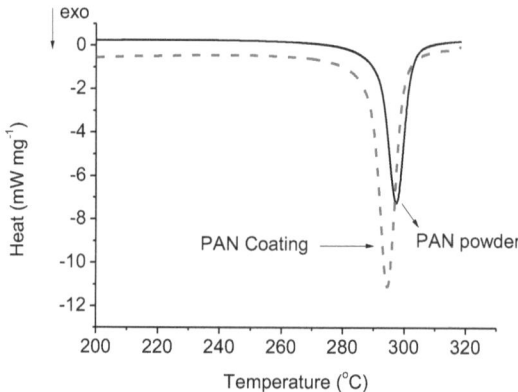

Figure 5.4.2 DSC thermogram of the powder and PAN coatings.

Neither shifts nor changes in signal intensity were observed for the nitrile signal comparing the PAN powder and the PAN coating. The same result was also observed in some investigations in the literature [4.40, 5.12] but the complex at the nitrile was still considered as the main interaction site. For instance, Phadke et al [4.40]. studied the interaction between DMF and PAN with different molecular weights and observed no changes for the nitrile peak. However he concluded that there was a complex between nitrile and solvent considering variations on the signals of pure DMF and DMF/PAN solutions. However, the shifts reported are usually in the range of 2 to 6 cm^{-1} and no information about the equipment resolution was given. Besides that, some shifts are reported in a range that goes from lower to higher values compared to the pure solvent making it complicate to draw any conclusions.

In the present study it is more likely that DMF interacts with other sites in the PAN molecule, as for example, the hydrogen at the tertiary carbon (methine carbon). In fact, another signal appears in the infrared spectrum of the coating in a frequency range that could be related to this interaction. However, these bands can also be related to the solvent. It was not possible to confirm this interaction by infrared spectroscopy. Some studies in the literature also discussed interactions between the methine hydrogen and DMF as discussed by Hattori et al. [5.13]. Hattori concluded that the interaction between isobutyronitrile (a model compound for PAN) and DMF occurs preferentially at the methine hydrogen instead of in the nitrile group. Therefore, due to the lack of evidence of an interaction between DMF and the nitrile group it can be assumed that DMF interacts with the methine hydrogen. This interaction is strong enough to maintaining certain amount of solvent even at temperatures as high as 280 °C as evidenced by the higher evolved heat during the cyclising of the coating.

In conclusion, it can be said that in a series of solvents that dissolves the polymer well, the one that is more volatile will produce more protective coatings. A similar trend is observed for PEI coatings, since that the more volatile solvent (DMAc) produced more protective coatings than the less volatile one (NMP). In PVDF this trend was not observed due to the lack of residual solvent in all prepared coatings. These results seem to be an obvious conclusion but there are some aspects that one should pay attention before selecting the more volatile solvent to use in the coating process. In general, the optimal solvent for the preparation of highly protective coatings should allow high polymer concentrations, dissolve well the polymer, be stable at room temperature and be easily removed.

A high polymer concentration is an important parameter to control the solution viscosity, and consequently, the final coating thickness. It was shown that the coating performance is extremely dependent on the coating thickness, especially when rough substrates are used. For PEI coatings, it was also commented that tests were done using CH_2Cl_2 (a very volatile solvent) solutions but bad performance were obtained, especially due to the low coating thickness which was a consequence of the low solution concentration.

The optimal solvent should also dissolve well the polymer otherwise the polymer will not be uniformly distributed on the substrate surface that will inevitably result in defects formation. A solvent that does not dissolve well the polymer will be probably instable at room temperature at certain concentrations, as is the case of PEI 10 wt.-% in DMF, This leads to complications to the coating process due to the occurrence of some processes like gelation. If a series of solvent satisfy these previous conditions the more volatile one should be selected since that it will result in coatings with lower residual solvent amount, otherwise, more drastic and expensive drying methods will be required.

5.4.2 – Influence of substrate pre-treatment

The influence of the substrate pre-treatment on the performance of PAN coatings was very similar to the one observed for PVDF. Both coatings showed very bad performance in ground, acetic acid cleaned and as-received substrate but good performance on HF-treated and nitric acid cleaned samples. The reasons for this are already discussed in the previous sections. The only marked difference between PAN and the other coatings is that a reaction between the polymer and the substrate surface take place even before the formation of corrosion products. It was shown in figure 4.2.6.7 that carbonyl signals appeared in the C 1s

XPS spectra at the interface, indicating a reaction between the hydroxides in the substrate surface with the nitrile group.

This reaction probably occurred during the drying of the samples due to the high temperature and could also be confirmed by infrared spectroscopy. While the CN/CH$_2$ ratio of the powder and the coating prepared on ground substrate was 0.86 this value dropped to 0.52 for the HF-treated substrate. Besides that, a new broad band appeared in the range of 1530 - 1600 cm^{-1}, shown in figure 5.4.3, which can be related to a cyclic structure. It can be seen in figure 5.4.3 that this broad band is not present in case of powder PAN and the coating on ground substrate. The literature shows that this cyclising reaction can occur in the presence of bases at temperatures as low as 25 °C that makes it very reasonable to considering its occurrence during the drying of the coatings [1.88]. The reaction mechanism is a mixture of the one presented in the first stage of figure 4.2.5.1 and the degradation mechanism which is shown in figure 5.4.1 where "X", in this case, is the hydroxide at the substrate surface.

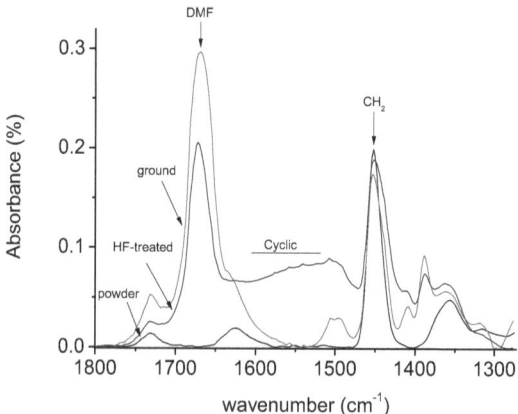

Figure 5.4.3 Infrared spectra of PAN coatings on ground and HF-treated substrate and of PAN powder.

This reaction is certainly responsible for the superior performance of PAN coatings on HF-treated substrate compared to others. Nevertheless, this reaction did not provided high adhesion to the system as can be observed in table 4.2.6.2. Besides that it should be pointed out that, comparing with PEI and PVDF, the aspect of the PAN coating on HF-treated substrate was the worse after the EIS tests. This indicates that the interface of the PAN HF-treated system was the less stable among the investigated ones. This can be related to the lower amount of polar groups in the polymer chain (compared to PEI) which results in poor adhesion, and to the presence of residual solvent that increased the diffusion of electrolytes. In case of PVDF, the lack of polar groups was compensated by the lack of residual solvent that provided good barrier properties. Nevertheless, the performance of PAN follows the same trend observed for PEI and PVDF in the sense that the coating on HF-treated substrate had the best performance due to interface reactions/interactions.

As was observed for the previous two polymers, no relation between substrate surface roughness and adhesion was obtained. This corroborates the conclusion made about PEI and PVDF that chemical interactions between substrate and polymer are the most important parameter to renders good adhesion. This is also the reason why the adhesion of PEI was much higher than that of PVDF and PAN since that PEI has much more polar groups in its structure allowing different interaction sites with the substrate. Another conclusion that is corroborated by all polymers is that pre-treatments that do not increase too much the substrate surface roughness are preferred, especially when the coating thickness is around 15 µm. Thin protective coatings require smooth cleaning substrate pre-treatments.

5.4.3 – Mechanism of coating degradation

As briefly discussed in section 4.2.6, PAN can also react with the formed magnesium hydroxide resulting in some derivatives as shown in figure 4.2.5.1 and figure 5.4.1. The hydrolysis of PAN is a well studied reaction although its exact mechanism is still under debate. The general scheme shown in figure 4.2.5.1 is a simplified mechanism of different possible reactions as discussed by Litmanovich [1.88]. The determination of the exact mechanism of hydrolysis is not the aim of this study, but rather the effect of this interfacial reaction on the coating performance. Therefore, only the aspects relevant for the coating behaviour will be considered here.

Aiming to investigate whether this interfacial reaction produced an adhesion increase, the coating over the exposed area was removed and infrared microscopy was used to analyze the damaged area. Similar to the observed for PVDF, a higher adhesion was observed during coating removal over the damaged area, indicating an increase in adhesion produced by the interfacial process. In figure 5.4.4 a microscopic view of the damaged substrate and infrared spectra of specific points of this sample are shown. It can be observed that the spectra are very similar to those shown in figure 5.3.3, for the PVDF coated HF-treated substrate after exposure. These similarities are expected since that both polymers have similar backbone structures (derivatives of polyethylene) and will produce similar hydrolysis products.

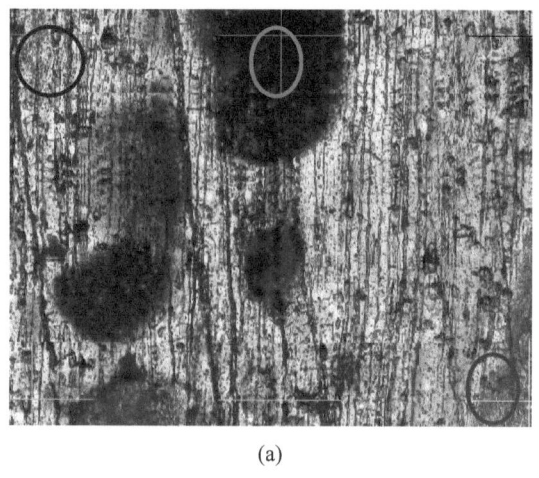

(a)

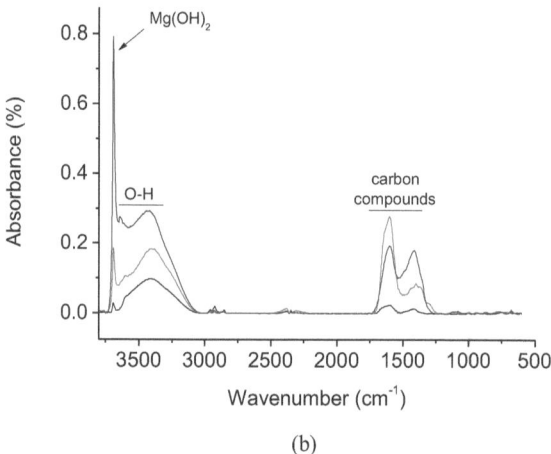

(b)

Figure 5.4.4: (a) Microscopic image of the damaged substrate after 2 months of exposure to a 3.5 wt.-% NaCl solution. (b) FT-IR spectra of specific points at the substrate surface.

The signals from 1250 to 1600 cm^{-1} are related to polymer derivatives, while the signals above 3000 cm^{-1} relate to O-H stretching which can be related to magnesium hydroxide and also to acidic derivatives of PAN. It is important to observe the lack of nitrile signal which indicates that no unmodified PAN has remained on the substrate. The lack of typical signals of CH_2 below 1000 cm^{-1} also corroborates this conclusion. It is also interesting to observe that the concentration of the polymer derivatives on the substrate surface changes from spot to spot reaching its maximum at the red spot. Comparing these spectra with the ones published in the literature [4.34-4.36] it can be concluded that these signals are related to cyclised PAN, as schematically shown in figure 5.4.1 for the thermal degradation mechanism. These spectra are especially similar to the one shown by Folcher et al.[4.34] when he reports the degradation mechanism of PAN at 210 °C. Folcher attributed the signals between 1530 to 1700 cm^{-1} to mixed vibration modes of C=N, C=C and N-H while the signals between 1250 and 1500 cm^{-1} were attributed to mixed CH_2, C-H, N-H, C-N, C-C vibration modes. Similar band assignments are presented by other researchers [4.35, 4.36].

Therefore, it can be concluded that cyclised derivatives of PAN are formed during the corrosion process and that it has stronger interaction with the substrate than the unmodified polymer. Besides that, the infrared and XPS spectra indicated the formation of carbonyl containing derivatives, as the ones shown in figure 4.2.5.1. Thus, both reactions probably occur at the interface. The acid site in the polymer derivative is probably the hydrogen linked to the nitrogen atom or even carbons between two nitrogen atoms, in case of the cyclised structure, and the carbonyl carbon or hydroxyl groups in case of derivatives shown in Figure 4.2.5.1.

5.4.4 – Potential use for biomedical applications

The tests made in simulated body fluid showed that some further improvements should be done before these coatings can be considered for biological applications. In general, the resistance obtained in immersion tests is not appropriate due to the low adhesion of the coating. Besides that, as DMF is an organic solvent with harmful effects to the human health, the coatings must have a much lower residual solvent amount to avoid intoxication. This requires drying processes more rigorous than the one applied in this study and maybe the replacement of DMF due to difficulties in completely remove it from PAN. Nevertheless, due to the interesting results published in the literature about PAN and even about HF-treated magnesium alloys in biological environments, the applications of PAN coated HF-treated substrates as biocompatible implants is an interesting area of research.

Besides that, a considerably high improvement in corrosion resistance of AZ31 alloys was produced by the HF-treatment followed by coating with PAN. It is also important to mention that AZ31 is an alloy with considerable high corrosion rate and it is expected that other alloys treated in the same manner will reach good stability in biological environments to be used as implants. To conclude this discussion it is important to remark once again that AZ31 is not an alloy suitable for biological applications due to allegations that aluminium may cause Alzheimer. Therefore, other alloying elements should be investigated for the preparation of magnesium implants. The results shown here servers to demonstrate the potential of the HF-treatment followed by PAN coating in the protection of the substrate.

6 – Summary and conclusions

Among the substrate pre-treatments investigated in this thesis, the one which renders the best performance for all the tested polymers is the HF one. This treatment decreases the Fe/Mn ratio to below its tolerance limit and creates a layer at the substrate surface, which contains mainly MgF_2 but also oxides and hydroxides. This layer can easily interact with polymers that contain certain acidity, and even react with them, as was observed for PAN. This allows the occurrence of interfacial reactions during the corrosion process which renders long-term stability for the coating. Besides that, a self-healing process was observed in case of PVDF coatings. This result shows that the creation of stable basic layers on the surface of magnesium alloys is an interesting approach for the development of high efficient pre-treatments for the application of polymer coatings.

The process which rendered the best performance was the HF-treatment followed by PEI dip-coating. The performance of the coatings treated in this manner was very good in impedance and immersion tests, and the coating showed good adhesion (5 MPa). PVDF coatings on the same substrate had also good performance but lower adhesion (2 MPa). The performance of PAN on HF-treated substrate was much inferior compared to the other two polymers. Among all the investigated coatings, the worst performance was observed for all spin-coated polymers in acetic acid cleaned substrates.

For all three investigated polymer systems, the pre-treatment that is more suitable for industrial applications is the cleaning by nitric acid. For PAN and PVDF the performance of the coatings on nitric acid cleaned samples was only inferior to that of the HF-treated ones. Nevertheless, for the spin-coating process other pre-treatments should be investigated as their performances in HF-treated and nitric acid cleaned substrates were much inferior to the observed for ground substrates. The cleaning of the samples with acetic acid resulted in the worse performance of all polymers and is not an appropriate pre-treatment for this application, especially, when thin coatings have to be prepared. For all polymers, no correlation between adhesion strength and substrate surface roughness was observed. The adhesion is much more related to interactions between polar groups of the polymer and the metal surface.

Both spin and dip-coating methods can produce highly protective PEI coatings when the substrate is adequately prepared. The best long-term impedance per coating thickness was obtained by spin-coated PEI. This show the potential use of PEI coatings in the corrosion protection of magnesium alloys. In case of PVDF the morphology of the coatings prepared by spin-coating was porous even when prepared under nitrogen atmosphere. The performance of spin-coated PAN was much inferior to the performance of dip-coated PAN due to the lower coating thickness.

For PEI and PAN, the solvent used for the coating preparation had considerable influence in the coating performance. This is due to the presence of residual solvent which enhanced the diffusion of electrolytes through the coating. The solvent also influenced the solution viscosity, and consequently, the final coating thickness. The best performance of PEI was observed when the coating was prepared using 20 wt. % solution in DMAc while for PAN the best performance was observed for a concentration of wt. 8% in DMF. In the case of PVDF coatings, no influence of solvents on the characteristics of the coatings was observed due to the low interaction between polymer and solvent which resulted in non-detectable residual solvent amount. The best performance of all coatings was observed at a coating thickness range of $11 - 15\mu m$ (considering a thickness variation at the vertical axis).

Treating the alloy with HF and coating it with PAN considerably increased the sample stability in simulated body fluid, compared to the untreated alloy. However, the samples showed considerable degradation after 21 days of immersion indicating that further improvements should be achieved to meet the requirements for biomedical applications. Nevertheless, AZ31 is an alloy with a considerable high corrosion rate and it is expected that the same treatment on a more stable magnesium alloy will produce sufficient stability to be used as an implant. The easiness in surface modification of PAN to induce biocompatibility, the beneficial effect of fluorides in the bone constitution and the high stability of this systems show that HF-treatment followed by coating with PAN is an interesting approach to prepare bio-implants with the required corrosion stability.

7 – Acknowledgements

I would like to express my gratitude to Dr. Nico Scharnagl for his assistance, corrections and suggestions during these three years of Ph.D.

I also would like to thanks Prof. Kainer and Prof. Wagner for their supervision as well as Prof. Reimers for being the Chairman of my thesis defence.

Especial thanks to the head of the "Korrosions und Oberflächetechnik" department of the Helmholz-Zentrum Geesthacht, Dr. Wolfgang Dietzel, for the many suggestions, English corrections and for always creating a nice and friendly work environment.

To all the colleagues at the WZK department of the Helmholz-Zentrum Geesthacht which gave me many suggestions and assistances during my work.

To all the friends and colleagues with I have spent three wonderful years.

To my parents, sisters, brother and niece as well as to my parents-in-law for everything that they have done to me, before and during this work.

To the almighty God for the many blessings which marked my live during these last three years.

And finally, to my beautiful wife Andrea for her love, support, and for the great moments that we shared during this time which were essential to my success. This work is dedicated to her who is the most important person in my life.

References

Chapter 1

1.1 http://minerals.usgs.gov/minerals/pubs/commodity/magnesium/
1.2 Kainer K. U., Srinivasan P. B., Blawert C., Dietzel W., "Corrosion of Magnesium and its alloys" Shreir's Corrosion, 2010, Chapter 3.09, pp 2011-2041.
1.3 Willekens J, „Magnesium-Verfügbarkeit, Markttendenzen Preisentwicklung, DGM Fortbildungsseminar", Clausthal-Zellerfeld, 29 (1997).
1.4 Kramer D. A., "Magnesium Metal" Annual report of the "United States Geographic Survey (USGS)" 2010.
1.5 Pinheiro G. A., "Local Reinforcement of Magnesium Components by Friction Processing: Determination of Bonding Mechanism and Assessment of Joint Properties", GKSS Forschunszentrum, 2008.
1.6 Yuen C. K., Ip W. Y., "Theoretical risk assessment of magnesium alloys as degradable biomedical implants" Acta Biomater. 6 (2010) pp. 1808-1812.
1.7 Witte F., Hort N., Vogt C, Cohen S, Kainer K. U., Willumeit R, Feyerabend F., "Degradable biomaterials based on magnesium corrosion" Curr. Opin. Sol. Stat. Mater. Sci. 12 (2008) pp. 63–72.
1.8 Ferreira P.C., Kde P. A., Takayanaguri A.M., Segura-Munoz A. I., "Aluminium as a risk factor for Alzheimers disease" Rev. Lat. Am. Enfermagem 16 (2008) pp 151-157.
1.9 Mordike B.L., Ebert T., "Magnesium: Properties - applications -potential" Mater. Sci. Eng. A302 (2001) pp 37-45.
1.10 United States Automotive Materials Partnership (USAMP) "Magnesium Vision 2020: A north American Automotive vision for magnesium" 2006.
1.11 Luo A. A. "Wrought magnesium alloys and manufacturing processes for automotive applications" in Lightweight Magnesium Technology, USA, 2006, pp 155-165.
1.12 Krajewski P. E., "Elevated Temperature Forming of Sheet Magnesium Alloys" Light Weight Magnesium Alloy Technology 2001-2005, 2006, pp. 331-336.
1.13 Schnell R., Hoenes R., Kaeumle F., „Erfahrungen mit Magnesium-Rädern in Sport und Serienautos" Deutscher Verband für Materialforschungs und Prüfung Korrosion an Fahrzeuge, DVM-Tag 1995, pp. 175-190.
1.14 Fink R., "Druckgießen von magnesium" in Magnesium-Eigenschaften, Anwendung, Potentiale, WILEY-VCH Verkag GmbH, Weiheim, 2000, pp 26-48.

1.15 Friedrich H., Schumann S., "Research for a "new age of magnesium" in the automotive industry", Mater. Proc. Tech. 117 (2001) pp 276-281.

1.16 Klein F., " Zukünftige potenziale von Magnesiumwerkstoffen in der automobileindustrie" in 15th Magnesium Automotive and User Seminar, Germany, 2007, pp 1-10.

1.17 Song L., Atrens A., "Corrosion mechanisms of magnesium alloys" Adv. Eng. Mater. 1 (1999) pp 11-33.

1.18 Davis J. R., "Corrosion: Understanding the Basics" ASM international, 4th Edition, 2008.

1.19 Witte F., "The history of biodegradable magnesium implants: A review", Acta Biomaterialia, 6 (2010) pp 1680-1692.

1.20 Lambotte A, L'utilisation du magnesium comme materiel perdu dans l'osteosynthese, Bull. Mem. Soc. Nat. Cir. 28 (1932) pp 1325-1334.

1.21 Wong H. Man, Yeung K. W. K., Lam K. O., Tam V., Chu P. K., Luk K. D. K., Cheung K. M. C., "A biodegradable polymer-based coatings to control the performance of magnesium alloy orthopaedic implants" Biomaterials, 31 (2010) pp 2084-2096

1.22 A. Pardo, M.C. Merino, A.E. Coy, R. Arrabal, F. Viejo, E. Matykina, "Corrosion behaviour of magnesium/aluminium alloys in 3.5 wt.% NaCl" Corros. Sci. 50 (2008) pp. 823- -834

1.23 Song G., Atrens A, Dargusch M., "Influence of microstructure on the corrosion of die cast AZ91D" Corros. Sci. 41 (1998) pp. 249-273.

1.24 Ambat A, Aung N., Zhou W., "Evaluation of microstructural effects on corrosion behaviour of AZ91D magnesium alloy" Corros. Sci. 42 (2000) pp. 1433-1455.

1.25 Lunder O., Lein J. E., Hesjevik S. M., "The role of Mg17Al12 phase in the corrosion of Mg alloy AZ91" Corrosion 45 (1989) 741-748.

1.26 Zeng R. C.., Zhang J., Huang W., Dietzel W., Kainer K. U., Blawert C., Ke W. "Review of studies on corrosion of magnesium alloys" Trans.Nonferrous Met. Soc. China 16(2006) pp. 763-771.

1.27 Liu M., Uggowitzer P. J.,, Nagasekhar A.V., Schmutz P., Easton M., Song G., Atrens A. "Calculated phase diagrams and the corrosion of die-cast Mg–Al alloys" Corros. Sci. 51 (2009) pp. 602–619.

1.28 Blawert C., Fechner D., Hoeche D., Heitmann V., Dietzel W. Kainer K. U., Zivanovic P., Scharf C., Ditze A., Groebner J., Schmid-Fetzer R.," Magnesium secondary alloys:

Alloy design for magnesium alloys with improved tolerance against impurities" Corros. Sci. 52 (2010) pp. 2452-2468.

1.29 Beldjoudi T., Fiaud C., Robbiola L., "Influence of homogenization and artificial aging heat treatments on corrosion behaviour of Mg-Al alloys" Corrosion 49 (1993) pp. 733-745.

1.30 Majumdara J. D., Galun R., Mordike B.L., Manna I., "Effect of laser melting on corrosion and wear resistance of a commercial magnesium alloy" Mater. Sci. Eng. 361 (2003) pp. 119-129.

1.31 Coy A.E., Viejo F., Garcia-Garcia F.J., Liu Z., Skeldon P., Thompson G.E. "Effect of eximer laser surface melting on the microstructure and corrosion performance of die cast AZ91D magnesium alloy" Corros. Sci. 52 (2010) pp. 387-397.

1.32 Guan Y.C., Zhou W., Li Z.L., Zheng H.Y., " Study on the solidification microstructure in AZ91D Mg alloy after laser surface melting" Appl. Surf. Sci. 255 (2009) pp. 8235-8238.

1.33 Nwaogu U.C., Blawert C., Scharnagl N., Dietzel W., Kainer K.U. "Efects of organic acid pickling on the corrosion resistance of magnesium alloy AZ31 sheet" Corros. Sci. 52 (2010) pp. 2143-2154.

1.34 Nwaogu U.C., Blawert C., Scharnagl N., Dietzel W., Kainer K.U. "Influence or inorganic acid pickling on the corrosion resistance of magnesium AZ31 sheet" Corros. Sci. 51 (2009) pp. 2544-2556.

1.35 Friedrich C. "Reliable light weight fastening of magnesium components in automotive applications" Light Weight Magnesium Alloy Technology 2001-2005, 2006, pp. 191-196.

1.36 Gray J. E., "Protective coatings on magnesium and its alloys – A critical review", J. Alloys Comp. 336 (2002) pp 88-113.

1.37 Winston A. W., Reid J. B., Gross W. H., "Surface preparation and magnesium painting of magnesium alloys" Ind. Eng. Chem. 27 (1935) pp 1333-1337.

1.38 Kouisni L., Azzi M., Zertoubi M., Dalard F., Maximovitch S. "Phosphate coatings on magnesium alloy AM60 part 1: study of the formation and the growth of zinc phosphate films" Surf. Coat. Tech. 185 (2004) pp. 58–67.

1.39 Zhou W., Shan D., Han E., Ke W., "Structure and formation mechanism of phosphate conversion coating on die-cast AZ91D magnesium alloy" Corros. Sci. 50 (2008) pp. 329–337.

1.40 Niu L.Y., Jiang Z.H., Li G.Y., Gu C.D., Lian J.S. "A study and application of zinc phosphate coating on AZ91D magnesium alloy" Surf. Coat. Tech., 200 (2006) pp. 3021-3026.

1.41 Eppensteiner F. W., Jenkins M. R., "Chromate conversion coatings" Metal finishing, 97 (1999) pp. 494-506.

1.42 Chiu K.Y., Wong M.H., Cheng F.T., Man H.C. "Characterization and corrosion studies of fluoride conversion coating on degradable Mg implants" Surf. Coat. Tech. 202 (2007) pp. 590-598.

1.43 Yan T, Tan L, Xiong D, Liu X., Zhang B., Yang K. "Fluoride treatment and in vitro corrosion behaviour of an AZ31B magnesium alloy" Mater. Sci. Eng. C 30 (2010) pp. 740–748.

1.44 Li Q, Zhong X, Hu J., Kang W. "Preparation and corrosion resistance studies of zirconia coating on fluorinated AZ91D magnesium alloy" Prog. Org. Coat. 63 (2008) pp. 222–227.

1.45 Thomann M., Krause C., Angrisani N., Bormann D., Hassel T., "Influence of a magnesium-fluoride coating of magnesium-based implants (MgCa0.8) on degradation in a rabbit model" J. Biomed. Mater. Res. 93A (2010) pp. 1609–1619.

1.46 Witte F., Fischer J. Nellesen J., Vogt C., Vogt J., Donath T., Beckmann F." In vivo corrosion and corrosion protection of magnesium alloy LAE442" Acta Biomaterialia 6 (2010) pp. 1792–1799.

1.47 Carboneras M., Hernández L.S.,. del Valle J.A, García-Alonso M.C., Escudero M.L." Corrosion protection of different environmentally friendly coatings on powder metallurgy magnesium" J. Alloys and Comp. 496 (2010) pp. 442–448.

1.48 Berglundh T., Abrahamsson I., Albouy J. P., " Bone healing at implants with a fluoride-modified surface: an experimental study in dogs" Clin. Oral Impl. Res. 18 (2007) pp. 147-152.

1.49 Lin C.S., Lin H.C., Lin K.M., Lai W.C. "Formation and properties of stannate conversion coatings on AZ61 magnesium alloys" Corros. Sci. 48, (2006) pp. 93-109.

1.50 Elsentriecy H. H., Azumi K, Konno H., "Improvement in stannate chemical conversion coatings on AZ91 D magnesium alloy using the potentiostatic technique" Electrochim. Acta, 53 (2007) pp. 1006-1012.

1.51 Ng, W.F. Wong M.H., Cheng F.T., "Cerium-based coating for enhancing the corrosion resistance of bio-degradable Mg implant" Mater. Chem. Phys., 119 (2010) pp. 384-388.

1.52 Wang C., Zhu S., Jiang F., Wang F., "Cerium conversion coatings for AZ91D magnesium alloy in ethanol solution and its corrosion resistance" Corros. Sci, 51 (2009) pp. 2916-2923.
1.53 Ghasemi A., Raja V.S., Blawert C., Dietzel W., Kainer K.U., "Study of the structure and corrosion behaviour of PEO coatings on AM50 magnesium alloy by electrochemical impedance spectroscopy" Surf. Coat. Tech. 202 (2008) pp. 3513-3518.
1.54 Arrabal R, Mathykina E., Hashimoto T., Skeldon P., Thompson G. E., "Characterization of AC PEO coatings on magnesium alloys" Surf. Coat. Tech. 203 (2009) 2207-2220.
1.55 Liang J., Srinivasan P. B., Blawert C., Dietzel W., "Influence of pH on the deterioration of plasma electrolytic oxidation coated AM50 magnesium alloy in NaCl solutions" Corros. Sci. 52 (2010) pp. 540-547.
1.56 Srinivasan P. B., Liang J., Blawert C., Störmer M., Dietzel W., "Characterization of calcium containing plasma electrolytic oxidation coatings on AM50 magnesium alloy" Appl. Surf. Sci. 256 (2010) pp. 4017-4022.
1.57 Liang J., Srinivasan P. B., Blawert C., Dietzel W "Comparison of electrochemical corrosion behaviour of MgO and ZrO_2 coatings on AM50 magnesium alloy formed by plasma electrolytic oxidation" Corros. Sci. 51 (2009) pp. 2483-2492.
1.58 Liang J., Srinivasan P. B., Blawert C., Störmer M., Dietzel W "Electrochemical corrosion behaviour of plasma electrolytic oxidation coatings on AM50 alloy formed in silicate and phosphate based electrolytes" Electrochim. Acta 54 (2009) pp. 3842-3850.
1.59 http://www.ahc-surface.com/files/mag_kepla_coat-engl.pdf
1.60 http://www.keronite.com/
1.61 http://www.tagnite.com/tagnite_coating/
1.62 Weiss K. D., "Paint and coatings: A mature industry in transition" Prog. Polym. Sci. 22 (1997) pp. 203-245.
1.63 Bierwagen G. P., "Film coating technologies and adhesion" Electrochim. Acta 37 (1992) pp. 1471-1478.
1.64 Bierwagen G. P., "Surface defects and surface flows in coatings" Prog. Org. Coat. 19 (1991) pp. 59-68.
1.65 Bierwagen G. P. "The physical chemistry of organic coatings revisited – viewing coatings as a materials scientist" J. Coat. Technol. Res. 5 (2008) pp. 133-155.

1.66 Sorensen P. A., Kiil S., Dam-Johansen K., Weinell C. E. "Anticorrosive coatings: a review" J. Coat. Technol. Res. 6 (2009) pp. 135-176.

1.67 Grundmeier G., Schmidt W. Stratmann M. "Corrosion protection by organic coatings: electrochemical mechanism and novel methods of investigation" Electrochim. Acta 45 (2000) pp. 2515-2533.

1.68 Scharnagl N., Blawert C., Dietzel W., "Corrosion protection of magnesium alloy AZ31 by coating with poly(ether imides) (PEI)" Surf. Coat. Technol. 203 (2009) 1423.

1.69 Jenekhe S. A., "The Rheology and spin-coating of polyimide solutions" Polym. Eng. Sci. 23 (1983) pp. 830-834.

1.70 Levinson W. A., Arnold A., Dehodgins O., "Spin-coating behaviour of polyimide precursor solution" Polym. Eng. Sci. 33 (1993) pp. 980-988.

1.71 Extrand C. W., "Spin-coating of very thin polymer films" Polym. Eng. Sci. 34 (1994) pp. 390-394.

1.72 Spangler L. L., Torkelson J. M., Royal J. S., "Influence of solvent and molecular weight on thickness and surface topography of spin-coated polymer films" Polym. Eng. Sci. 30 (1990) pp. 644-652.

1.73 Gupta S. A., Gupta R. K., "A parametric study of spin-coating over topography" Ind. Eng. Chem. Res. 37 (1998) pp. 2223-2227.

1.74 Eisenbraun et al. "Production of electronic coatings a partially fluorinated polyamic acid composition" US Patent 4996254.

1.75 Eisenbraun et al. "Production of electronic coatings by spin-coating a partially fluorinated polyimide composition" US 4997869.

1.76 Siau S., Vervaet A., Degrande S., Schacht E., Calster A. V., "Dip-coating of dielectric and solder mask epoxy polymer layers for build-up purposes" Appl. Surf. Sci. 245 (2005) pp. 353-368.

1.77 Fang H., Li K., Su T., Yang T. C., Chang J., Lin P., Chang W., "Dip-coating assisted polylactic acid deposition on steel surface: Film thickness affected by drag force and gravity" Mater. Let. 62 (2008) pp. 3739-3741.

1.78 Yimsiri P., Mackley M. R., "Spin and dip-coating off light-emitting polymer solutions: Matching experiment with modeling" Chem. Eng. Sci. 61 (2006) pp. 3496-3505.

1.79 Yu B., Uan J., " Sacrificial Mg film anode for cathodic protection of die cast Mg-9 wt.%Al-1 wt.%Zn alloy in NaCl aqueous solution" Script. Mater. 54 (2006) pp. 1253-1257.
1.80 Lu X., Zuo Y., Zhao X., Tang Y., Feng X., " The study of a Mg-rich epoxy primer for protection of AZ91D magnesium alloy" Corros. Sci. Article in press.
1.81 Ross G. J., Watts J. F., Hill M. P., Morrissey P., "Surface modification of poly(vinylidene fluoride) by alkaline treatment 1. The degradation mechanism" Polymer 41 (2000) pp. 1685-1696.
1.82 Vecellio M. "Opportunities and developments in fluoropolymeric coatings" Prog. Org. Coat. 40 (2000) pp. 225-242.
1.83 Leivo E., Wilenius T., Kinos T., Vuoristo P., Mantyla T., "Properties of themally sprayed fluoropolymer PVDF, ECTFE, PFA and FEP coatings" Prog. Prg. Coat. 49 (2004) pp. 69-73.
1.84 Verdianz T., Simburger H., Liska R., "Surface modification of imide containing polymers I: Catalytic groups" Europ. Pol. J. 42 (2006) pp. 638-654.
1.85 Lee K. W., Kowalczyk S. P., Shaw J. M., "Surface modification of PMDA-oxydianiline polyimide. Surface structure-adhesion relationship" Macromol. 23 (1990) pp. 2097-2100.
1.86 Thomas R. R., Buchwalter S. L., Buchwalter L. P., Chao T. H. "Organic chemistry on a polyimide surface" Macromol. 25 (1992) pp. 4559-4568.
1.87 Lee K. W., Kowalczyk S. P., Shaw J. M., "Surface modification of BPDA-PDA polyimide" Langmuir 7 (1991) pp. 2450-2453.
1.88 Litmanovich A. D., Plate N. A., "Alkaline hydrolisis of polyacrylonitrile. On the reaction mechanism" Macromol. Chem. Phys. 201 (2000) pp. 2176-2180.
1.89 Wang Z., Wan L., Xu Z., "Surface engineerings of polyacrylonitrile-based asymmetric membranes towards biomedical applications: An overview" J. Membr. Sci. 304 (2007) pp. 8-23.
1.90 Barnartt S. "Electrochemical nature of corrosion" Electrochemical techniques for corrosion engineering (1986) pp. 1-11.
1.91 Hack H. P., "The potentiostatic technique for corrosion studies" Electrochemical techniques for corrosion engineering (1986) pp. 57-65.
1.92 Shi Z., Liu M., Atrens A., "Measurements of the corrosion rate of magnesium alloys using Tafel extrapolation" Corros. Sci. 52 (2010) pp 579-588.

1.93 Song G., Atrens A., Stjohn D., Nairn J., Li Y., " The electrochemical corrosion of pure magnesium in 1N NaCl" Corros. Sci. 39 (1997) pp. 855-875.

1.94 Thomaz T. R., Weber C. R., Pelegrini T., Dick L. F. P., Knornschild G., "The negative difference effect of magnesium and of the AZ91 alloy in chloride and stannate containing solutions" Corros. Sci. 52 (2010) pp. 2235-2243.

1.95 Grundmeier G., Simoes, A. " Corrosion protection by organic coatings" in Encyclopedia of Electrochemistry, Vol.4, WILEY-VCH Verlag GmbH, Weinheim, 2003, 500.

1.96 Silverman D. C. "Primer on the AC impedance technique" Electrochemical techniques for corrosion engineering (1986) pp. 73-79.

1.97 Westing E. P. M., Ferrari G. M., Wrr J. H. W., "The determination of coating performance with impedance measurements - II. Water uptake of coatings" corros. Sci. 36 (1994) pp. 957-977.

1.98 Belluci F., Nicodemo L., "Water transport in organic coatings" Corrosion 49 (1993) pp.235-247.

1.99 Lindqvist S. A., "Theory of dielectric properties of heterogeneous substances applied to water in a paint film" Corrosion 41 (1985).

1.100 Titz J., Wagner G. H., Spahn H., Ebert M., Juttner K., Lorenz W. J., "Characterization of organic coatings on metal substrates by electrochemical impedance spectroscopy". Corrosion 46 (1990) pp. 221-228.

Chapter 4

4.1 Revie R.W., Uhlig's Corrosion Handbook, John Wiley & Sons, 2 Edition, 2000.

4.2 Hanawalt J.D., Nelson C.E., Peloubet J.A, "Corrosion studies of magnesium and its alloys" Trans. AIME. 147 (1942) pp 273-299.

4.3 Loose W.S., "Corrosion and Protection of Magnesium". ASM Int, Materials Park, OH 1946, pp 173-260

4.4 Zamin M., "The Role of Mn in the corrosion behaviour of Al-Mn Alloy" Corrosion 37 (1981) pp. 627-632.

4.5 Reichek K.N., Clark K.J., Hillis J.E., "Controlling the salt Water Corrosion Performance of Magnesium AZ91 alloy", SAE Technical Paper series 850 417 (1985).

4.6   Makar G.L., Kruger J., "Corrosion of magnesium" Int. Mater. Rev. 38 (1993) pp. 138-153.
4.7   Kloprogge J.T., Hickey L., Frost R.L., "FT-Raman and FT-IR spectroscopic study of synthetic Mg/Zn/Al-hydrotalcites" J. Raman Spectrosc. 35 (2004) pp. 967-974.
4.8   Lide R.D., CRC Handbook of Chemistry and Physics – 84th, 2003.
4.9   Jönsen M., Persson D., Thierry D., "Corrosion product formation during NaCl induced atmospheric corrosion of magnesium alloy AZ91D" Corros. Sci. 49 (2007) pp. 1540-1558.
4.10 Karmakar B., Kundu P., Dwivedi R.N., "IR spectra and their application for evaluating physical properties of fluorophosphates glasses" J. Non-Cryst. Solids 289 (2001) pp. 155-162.
4.11 Wolf W.L., The Infrared Handbook, Washington, Office of naval research, 1978.
4.12 Morterra C., Cerrato G., Cuzzato P., Masiero A., Padovan M., " Infrared surface characterization of AlF_3" J. Chem. Soc. Faraday Trans. 88 (15) (1992) pp. 2239-2250.
4.13 Prescott H.A., Li Z., Kemnitz E., J, Deutsch, H. Lieske, "New magnesium oxide fluorides with hydroxyl groups as catalysts for Michael additions" J. Mater. Chem. 15(2005) pp. 4616-4628.
4.14 Flack W.W., Soong D.S., Bell A.T., Hess D.W., "A mathematical model for spin-coating of polymer resists" J. Appl. Phys. 56, (1984) pp. 1199-1206.
4.15 Jenekhe S.A., Schuldt S.B., "Coating flow of non-Newtonian fluids on a flat rotating disk" Ind. Eng. Chem. Fundam. 23 (1984) pp. 432-436
4.16 Jasinski P, Molin S, Gazda M, Petrovskyc V, Anderson H U, "Applications of spin-coating of polymer precursor and slurry suspensions for Solid Oxide Fuel Cell fabrication" J. Power Sources, 194 (2009) pp. 10-15.
4.17 Chena K., Tiana Y., Lüa Z., Aia N., Huanga X., Su W., "Behavior of 3 mol% yttria-stabilized tetragonal zirconia polycrystal film prepared by slurry spin-coating" J. Power Sources, 186 (2009) pp. 128-132
4.18 Mulder M., "Basic Principles of Membrane Technology", Kluwer Academic Publishers, Netherlands, 1991.
4.19 D. Wang, K. Li, W.K. Teo, "Phase separation in polyetherimide/solvent/nonsolvent systems and membrane formation" J. Appl. Polym. Sci. 71 (1999) pp. 1789-1796.
4.20 Skrobis K.J., Denton D.D., Skrobis A.V, " Effect of early solvent evaporation on the mechanism of spin-coating of polymeric solutions" Poly. Eng. Sci. 30 (1990) pp. 193-196.

4.21 Kurdi J., Kumar A., "Structuring and characterization of a novel highly microporous PEI/BMI semi-interpenetrating polymer network" Polymer, 46 (2005) pp. 6910-6922.

4.22 Cheng H.L., You J., Porter R.S., "Intermolecular interaction and conformation in poly(ether ether ketone)/poly(ether imide) blends – An infrared spectroscopic investigation" J. Polym. Res. 3 (1996) pp. 151–158.

4.23 Kostina Y V, Bondarenko G N, Alent'ev A Y, Yampol'skii Y P, "Effect of structure and conformational composition on the transport behaviour of poly(ether imides)" Polymer Science, Ser. A 49 (2007) pp. 77-88.

4.24 Barzin J., Sadatnia B., "Theoretical phase diagram calculation and membrane morphology evaluation for water/solvent/polyethersulfone systems" Polymer 48 (2007) pp. 1620-1631.

4.25 Gelfandbein V, Perlman M. M., "Substrate effects on crystallization of polyvinylidene fluoride from solution" J. Mater. Sci. 18 (1983) pp. 3183-3189.

4.26 Gregorio R, Cestari M, "Effect of crystallization temperature on the crystalline phase content and morphology of poly(vinylidene fluoride)" J. Polym. Sci. B 32 (1994) pp. 859-870.

4.27 Brandrup J., Immergut E. H., Grulke E. A., Polymer Handbook, 4th Edition, John Willey & Sons.

4.28 Bormashenko Y., Pogreb R., Stanevsky O., Bormashenko E., "Vibrational spectrum of PVDF and its interpretation" Polym. Test. 23 (2004) pp. 791–796.

4.29 Boccaccio T., Bottino A., Capannelli G., Piaggio P., "Characterization of PVDF membranes by vibrational spectroscopy" J. Memb. Sci. 210 (2002) pp. 315–329.

4.30 Kobayashi M., Tashiro K., Tadokoro H., "Molecular vibrations of three crystal forms of poly(vinylidene fluoride)" Macromol. 8 (1975) pp. 158-171.

4.31 Wojciechowska M., Zielinsk M., Pietrowski M., "MgF_2 as a non-conventional catalyst support" J. Fluor. Chem. 120 (2003) pp. 1–11.

4.32 Yang M.C., Tong J.H., "Loose ultrafiltration of proteins using hydrolyzed polyacrylonitrile nanofiber. J. Memb. Sci. 132 (1997) pp. 63-71.

4.33 Godjevargova T., Dinamov A., "Permeability protein adsorption of modified charged acrylonitrile copolymer membranes." J. Memb. Sci. 67 (1992) pp.283-287.

4.34 Folcher H.S., Mooney J.R., Ball L.E., Boyer R.D., Grasselli J.G., "Infrared and NMR spectroscopic studies of thermal degradation of polyacrylonitrile) Spectrochim. Acta 41 (1985) pp. 271-278.

4.35 Xue T.J., McKinney M.A., Wilkie C. A., "The thermal degradation of polyacrylonitrile" Pol. Deg. Stab. 58 (1997) pp.193-202.

4.36 Deng S., Bai R., Cheng J.P, "Behaviors and mechanism of copper adsorption on hydrolyzed polyacrylonitriles fibers" J. Col. Inter. Sci. 260 (2003) pp 265-272.

4.37 Scharnagl N., Buschatz H. "Polyacrylonitrile (PAN) membranes for ultra and microfiltration" Desalination 139 (2001) pp. 191-198.

4.38 Chen J. C., Harrison I. R., "Modification of polyacrylonitrile (PAN) carbon fiber precursor via post-spinning plasticization and stretching in dimethyl formamide (DMF)" Carbon 40 (2002) pp. 25-45.

4.39 Sanchez-Soto P. J., Aviles M. A., del Rio J. C., Gines J. M., Pascual J., Perez-Rodriguez J. L., "Thermal study of the effect of several solvents on the polymerization of acrylonitrile and their subsequent pyrolysis" J. Anali. Appl. Pyrol. 58-59 (2001) pp. 155-172.

4.40 Phadke M. A., Musale D. A., Kulkarni S. S., Karode S. K., "Poly(acrylonitrile) ultrafiltration membranes. I. Polymer-salt-solvent interactions" J. Polym. Sci. B 43 (2005) pp. 2061-2073.

4.41 Frauchiger L., Taborelli M., Aronsson B. O., Descouts P., "Ion adsorption on titanium surfaces exposed to a physiological solution" Appl. Surf. Sci. 143 (1999) pp. 67-77.

Chapter 5

5.1 Verdier S., Laak N., Delalande S., Metson J., Dalard F., "The surface reactivity of a magnesium–aluminium alloy in acidic fluoride solutions studied by electrochemical techniques and XPS" Appl. Surf. Sci. 235 (2004) pp. 513-524.

5.2 Booster J.L., Sandwijk A.V., Reuter M.A., "Conversion of magnesium fluoride to magnesium hydroxide" Miner. Eng. 16 (2003) pp. 273-281.

5.3 Momber A. W., Koller S., Dittmers H. J., "Effect of surface preparation methods on adhesion of organic coatings on steel substrates" J. Protec. Coat. Lin. November (2004) pp. 44-50.

5.4 Momber A. W., Greverath W. D., "Surface Preparation standards for steel substrates" J. Protec. Coat. Lin. February (2004) pp. 48-52.

5.5 Hatton J. V., Richards R. E., " Solvent effects in the proton resonance spectra of dimethyl-formamide and dimethyl-acetamide" Mol. Phys. 3 (1960) pp. 253-263.

5.6 Strunskus T., Grunze M., Kochendoefer G., Woll Ch., "Identification of physical and chemical interaction mechanisms for the metals gold, silver, copper, palladium, chromium, and potassium with polyimide surfaces" Langmuir 12 (1996) pp. 2712-2725.

5.7 Ramos M. M. D., "Theoretical study of metal-polyimide interfacial properties" Vacuum 64 (2002) pp. 255-260.

5.8 Lando J., Olf M., Peterlin A., "Nuclear magnetic resonance and x-ray determination of the structure of poly(vinylidene fluoride)" J. Polym. Sci. Part A 4 (1966) pp. 941-951.

5.9 Brewis D.M., Mathieson I., Sutherland I., "Pretreatment of poly(vinyl fluoride) and poly(vinylidene fluoride) with potassium hydroxide" Int. J. Adhes. Adhes. 16 (1996) 87–95.

5.10 Ross G.J., Watts J.F., Hill M.P., Morrisey P., "Surface modification of poly(vinylidene fluoride) by alkaline treatment1. The degradation mechanism" Polymer 41 (2000) pp. 1685–1696.

5.11 Voinkova I.V., Ginchitski N.N., Gribov I.V., Klebanov I.I., Kuznetsov V.L., Moskvina N.A., Pesin L.A., Evsyukov S.E., " A model of radiation-induced degradation of the poly(vinylidene fluoride) surface during XPS measurements" Polym. Deg. Stab. 89 (2005) pp. 471–477.

5.12 Minagawa M., Shirai H., Morita T., Fujikura Y., Kameda Y., "Dynamic nuclear magnetic resonance and Raman spectroscopic measurements of five kinds of N,N-dimenthylformamide derivatives in relation to the dissolution mechanism of polyacrylonitrile" Polymer 37 (1996) pp. 2353-2358.

5.13 Hattori M., Yamazaki H., Saito M., Hisatani K., Okajima K., "NMR study on the dissolved state of polyacrylonitrile in various solvents" Polym. J. 28 (1996) pp. 594-600.

Die VDM Verlagsservicegesellschaft sucht für wissenschaftliche Verlage abgeschlossene und herausragende

Dissertationen, Habilitationen, Diplomarbeiten, Master Theses, Magisterarbeiten usw.

für die kostenlose Publikation als Fachbuch.

Sie verfügen über eine Arbeit, die hohen inhaltlichen und formalen Ansprüchen genügt, und haben Interesse an einer honorarvergüteten Publikation?

Dann senden Sie bitte erste Informationen über sich und Ihre Arbeit per Email an *info@vdm-vsg.de*.

Sie erhalten kurzfristig unser Feedback!

VDM Verlagsservicegesellschaft mbH
Dudweiler Landstr. 99
D - 66123 Saarbrücken
www.vdm-vsg.de

Telefon +49 681 3720 174
Fax +49 681 3720 1749

Die VDM Verlagsservicegesellschaft mbH vertritt

Printed by Books on Demand GmbH, Norderstedt / Germany